DANCING IN THE DARK

THE "WALTZ IN WONDER" OF QUANTUM METAPHYSICS

DANCING IN THE DARK

THE "WALTZ IN WONDER" OF QUANTUM METAPHYSICS

By
Ronald Keast, PhD

iUniverse, Inc.
New York Bloomington

Dancing in the Dark
The "Waltz in Wonder" of Quantum Metaphysics

iUniverse books may be ordered through booksellers or by contacting:

iUniverse
1663 Liberty Drive
Bloomington, IN 47403
www.iuniverse.com
1-800-Authors (1-800-288-4677)

ISBN: 978-1-4401-3695-5 (sc)
ISBN: 978-1-4401-3697-9 (hc)
ISBN: 978-1-4401-3696-2 (ebook)

Library of Congress Control Number: 2009925677

Printed in the United States of America

iUniverse rev. date: 6/25/2009

Contents

Acknowledgments

Sincere thanks to my friend and former colleague, Dr. Howard Barrows, for his support and very helpful suggestions during the writing of this book; to my daughter, Tabitha, for some practical copyediting assistance; and above all, to my wife, Elizabeth, who has been loving and supportive in this as in everything I've done and tried to do.

Introduction

The "Waltz in Wonder" of Quantum Metaphysics

A song written for a Broadway show in the 1930s has remained, if not exactly popular, certainly a classic to this day. The reason, I suspect, has as much to do with the lyrics as with the music. The lyrics reflect some of our deepest thoughts and fears; that life is a dance in the dark and that it soon ends; in the meantime, we waltz in the wonder of why we are here, while life hurries by. The name of the song is "Dancing In The Dark."[1]

If we think about it—and most of us think about it from time to time—it is obvious that we are all dancing in the dark, while only some of us waltz in wonder. Where did we come from? Why are we here? Where are we going? What is reality? What are we, and the universe, made of? What is true? Is there such a thing as truth? Can we know what it is? And, perhaps the most fundamental question of all, why is there something rather than nothing? Philosophers, theologians, scientists, and of course, artists and writers of all stripes have been speculating about these and other such questions forever. "The unrest which keeps the never stopping clock of metaphysics going is the thought that the non-existence of the world is just as possible as its existence."[2]

The question for this writer and this book is who is currently doing the most interesting, and perhaps most relevant, speculation about such issues and what effects such speculation might have. Who are the most interesting metaphysicians? Is it the traditional theologians, connected with formal religious organizations? Hardly! Their angels are still dancing on the heads of their various pins. Is it the professional philosophers, connected with formal university departments? I don't think so! One doesn't hear much about philosophers and philosophy departments anymore. Is it the "popular" writers whose books continue to denude our forests each year? Not likely! Serious thought and speculation about such issues can be deemed religious and, therefore, politically incorrect by many.

While I'm sure some speculation must emanate from some of these and other such sources, my vote goes to the scientists, specifically to physicists and the theoretical physicists who immerse themselves in the theories of quantum mechanics. This may be because I have come comparatively late to these writings and so have become enthralled with them in a way that I have not been for years by writings from other sources.

The questions raised by these physicists about truth and reality are so utterly fascinating and so utterly profound that modern physicists have returned science to its original role as natural philosophy and to its roots in metaphysics.

From my perspective, quantum physics may be better described as quantum metaphysics. Certainly, the quantum theories proposed by theoretical physicists have profound metaphysical and theological implications.

This is a good thing.

Physics, and science generally during its long history as natural philosophy, has always included major elements of metaphysics, which is defined by *Webster's New World Dictionary* as a "division of philosophy that is concerned with the fundamental nature of reality and being, and that includes ontology (the study of what is outside objective experience), cosmology (the study of the very large, i.e., the universe), and often epistemology (the study of language and meaning)." The more specific definition of metaphysical as "of or relating to the transcendent or to a reality beyond what is perceptible to the senses" is precisely what quantum mechanics deals with. Its theories are wonderfully speculative and now influence scientific thought about all of reality from the infinitely small to the infinitely large to everything in between.

Faith in and the search for reality and truth have been and remain the common waltz in wonder of philosophy, theology, mysticism, and science. The metaphors for the ultimate goal vary, as do the methods of the waltz, but the goal is the same. It is to know, to verily know. Truth and reality—in an ultimate sense, with a capital

T and a capital *R* if you will—can be understood as metaphors for God just as God can be understood as a metaphor for ultimate Truth and Reality. To know one is to know the other. It is in this sense that the historic and very current search of science, specifically of quantum physics, for reality and truth can be seen as a search for God. Likewise, the theoretical physicists leading this search may be described not only as metaphysicians but also as theologians; theology is knowledge of God, and knowledge of truth and reality is just that.

This deep metaphysical speculation has been an important part of science in the West since the beginning of natural philosophy in ancient Greece. The Greeks were interested in determining the fundamental elements or nature of the universe, whether this was considered matter or form. The reality of and truth about the universe was clearly elucidated by Isaac Newton, who believed that, outside objective experience, God was responsible for starting it all and for setting the rules. Newton's mathematical laws about the way the universe worked were originally thought of as accessing the mind of God. However, this quaint nod to religious orthodoxy was soon found to be unnecessary. The laws just existed.

Newton's scientific truths have been the basis of classical physics and the predominant popular view of reality for over three hundred years. These scientific truths have provided the foundation for our modern society, with its central myth of science and progress. They have been the basis of the secular religion of science in the West. However, these truths have now been overturned—first by Einstein's theories of relativity and then by the theories of quantum mechanics.

Many of these theories and speculations are far stranger than the strangest science fiction, and they can lead the unwary and unreflective to purely fantasy speculations. This does not do justice to the kind of intellectual rigor and critical examination that has gone into and continues to go into these theories.

Besides the content, it is this breadth and depth of quality of thought, with its constant examination and eventually, where possible, experimentation, that makes science, and especially the new science, such a rewarding field of study, even for a nonscientist. Obviously, even the new physics does not deal with all, or even most, of the important human questions. It says nothing about love or feelings or human relations or most of the issues we deal with from day to day. Traditionally, it does not take into account or think much about human consciousness. However, what it does deal with, and the way in which it deals with it, is more than enough to make it, arguably, the most interesting and far and away the most speculative part of philosophy today. And since philosophy is the love of wisdom, all such lovers should take an interest.

In addition, based on the Webster definition, we are all metaphysicians. All of us, at least some of the time, think about these issues. Many do so from a particular religious perspective. Many do so from a secular, nonreligious perspective. But we all do so from some belief perspective. Nobody escapes this. As we will see, the hunger for reality, or truth, is indeed the mother of all metaphysics. This hunger motivates science, philosophy, and religion. Our peculiar faiths and our beliefs are all colored by our circumstances, by the very fact that we are human and live in a certain environment and have been subject to specific and varied psychological, physical, and spiritual influences. We all believe in something, even if we believe it to be nothing.

A belief that we are children of God and that Jesus Christ is the savior of the world, is a belief; there is no proof, or certainly none in the classical scientific sense of repeatable experimentation or mathematical formulae. The belief that we are all that is, that there is nothing more, that the idea of truth and of anything beyond or behind our physical circumstance is illusion, that "life is a bitch, and then you die," is still a belief, a faith; there is no proof for this either.

Quantum mechanics is the science of the very tiniest, subatomic level of reality. This is the level of so-called elementary particles, whether these are particles, quanta, mass, and/or energy. This is the basic, fundamental level of reality on which the reality that we see around us is built, that of our sensory experience. Mechanics is a seemingly strange term, but its definition makes the usage clear: "The branch of physics that deals with motion and the phenomena of the action of forces on bodies" (*Webster's New World Dictionary*). The science consists of mathematical formulae for picturing the nature and the behavior of these elementary particles, the motion and phenomena of energy and particles. The science builds a description, a mathematical metaphor, for reality and thus for truth.

It is generally understood that language—words as metaphor—is not adequate to communicate reality or truth and that it is always approximate. This is one reason why mathematics is preferred to language. However, while mathematics is more accurate, its realities and truths often do not translate well into words; in fact, they have no obligation to translate into words.

Many prominent physicists, as well as some very accomplished science writers, are trying to put into words the extraordinary findings and mathematical theories of the new physics. They are also trying to evolve laboratory experiments that will prove, in the scientifically traditional sense of the term, at least some of the theories. This has proven to be very difficult—in most cases, impossible.

There are few things that can be scientifically proven. A mathematical theory or equation is one, but it must only be internally consistent, with no obligation to correspond to an external physical reality. Nobel Laureate physicist Robert Laughlin says that "deep down inside every physical scientist is the belief that measurement accuracy is the only fail-safe means of distinguishing what is true from what one imagines, and even of defining what true means."[3]

I have real sympathy for this view of truth, limited and truncated as it obviously is. Experimental science has been the cornerstone of secular religion in the

West. It relieves those who accept it, at least in their own minds, from dancing in the dark. The truths of experimental science are objective and open to unlimited verification and falsification. Quantum mechanics has gone far beyond any hope of this sort of simple verification. Indeed, one of the most basic quantum theories undermines the very nature of objectivity itself, stating that the experimenter can never be objectively separated from the experiment. It also challenges the most sacred tenet of Western secularism—the truths of science. This subjectivity has caused a major crisis in theoretical physics.

The nature of the "religion of science" was beautifully summarized in an influential article written in 1958 by the French philosopher Jacques Ellul and published in the American magazine *Diogenes*. The article, called "Modern Myths," identified the central myths, or religious beliefs, of the modern age in the West as science and technological progress. Ellul furthered this thesis in a book called *The Technological Society* and other works. He defined myth as, in part, "the expression of profound, permanent tendencies" and said that "myth always includes an element of belief, of religious adherence, of the irrational, without which it could never express on behalf of man what it was supposed to convey."[4]

The science that Ellul referenced, which has provided the fundamental secular religion of the modern age, was that of classical physics, or the science of Newton. This is the commonsense science that the majority of us still believe in. It is the science that explains the truths of our world—that matter, the "stuff" around us, is real and solid; that space is open and empty and separates matter; that time is linear, separate and distinct; that reality has three dimensions: up and down, side to side, forward and backward.

However, this classical science and all of these truths have been weakened by both relativity and quantum mechanics and proven to be, while not strictly untrue in a limited sense because they are still adequate for explaining everyday things, inadequate to explain or express the full nature of reality or truth. What we all take for granted as real is not real, or at least is not the whole of reality. There is a reality, perhaps a myriad of realities, beyond our perception and beyond what, up to this time at least, has been our popular intellectual understanding.

The entire edifice of secular truths and the metaphors that we use to express them have thus been undermined, and the new science has not provided new universal truths and metaphors to capture the mind the way the old ones did and still do, at least not yet. Rather, it has opened up question after question, doubt after doubt, showing us a world of uncertainties. But, just as many traditional religious believers were little swayed over hundreds of years by the old science, modern secular believers are not being swayed by the deaths of their scientific gods. They should be.

Relativity and quantum mechanics have already created a revolution in science. The entire bulwark of scientific truths that have held the minds of millions of deep thinkers and much of the general public for the past hundreds of years—truths that themselves challenged and, in many ways, destroyed the religious truths of the hundreds of years before that—have been utterly undermined by the findings and understandings of the new physics.

Nobody except primarily the scientists and a few interested observers knows about it. This is a shame for anybody who thinks and wonders and questions but has no intelligent outlet for such creative activity within either his particular sacred religious community or his secular religious community. If it is true that it is not truth but the pursuit of truth that makes us free—and I for one believe that it is—then the fact that this quite extraordinary pursuit is happening below the radar of popular consciousness may be dangerous for the future of our democracy.

Theoretical physics has uncovered a reality of uncertainty that underlies the world and reality that we know, setting the ground rules for science in general and thus for modern secular thought. The old certainties are gone. One can ignore their disappearance and maintain one's old certainties, either secular or sacred. But this would surely be a delusion, as well as an intellectual, emotional, and spiritual loss, because uncertainty, while it can be unsettling to the lazy mind, does not contradict or undermine faith. In fact, it can provide a newer and vastly more energizing ground for faith than the old certainties could. In turn, faith can provide assurance, with or without certainty.

Quantum theory provides the basis for a modern

reformation. It has undermined the old secular religion and the old secular certainties. But, in so doing, it has provided a central position for faith, perhaps for faith alone.

The reformers, the quantum theorists, are continuing a long philosophic tradition in the West of a continuous and reasoned pursuit of truth and the questioning of all truths. What is important to remember, and what is not always clear in their writings, is that they, like the scientists and other philosophers before them, have not discovered truth but are pursuing it in the faith that it is there and worth pursuing. It may be called a "theory of everything," a "string theory," or "the theory of emergent properties," but it is a theory that will always be open for debate and for debunking. It is undoubtedly true that "no theory can ever explain why anything is—that is the supreme mystery. But theory may be able to tell us why one thing rather than another is created and experienced."[5]

What is equally important to remember, especially about these current theories, is that nobody—and I mean nobody, including the physicists themselves and most certainly this writer—"gets" the quantum reality. It has been, still is, and perhaps always will be a mystery.

So, what is reality, what is true, what is God? Can we know these things?

To be just a tad irreverent, as the town preacher so eloquently puts it in the wonderful movie *Blazing Saddles*, "Do we have the strength to carry on this mighty task…? Or are we all just jerking off?" The great theologian St. Augustine defined the belief that we can know what God is as "*fantastica fornicatio*," mind masturbation, or fornicating with our own fantasies. The same thing could be said of a belief that we can know what truth or reality is.

Ambrose Bierce's *The Devil's Dictionary* defines reality as "the dream of a mad philosopher."[6] In this case, most philosophers—natural and otherwise—as well as most theologians, are mad. However, the pursuit of truth or reality or God—in the faith that such pursuit is worthwhile—is a divine madness, a madness that is a basis for intellectual and spiritual knowledge and progress.

While we may be dancing in the dark when we chase our understanding of these words, it is a waltz in wonder. So, let's learn some elementary steps and join the dance.

Chapter One
The Unreal and Uncertain Reality of Quantum Mechanics

The commonsense view of the world, the one that most of us share, is that reality is revealed to us through our senses. Some believe that this is all there is to it; others that there is more to reality than this. This "more" may be what our traditional religions teach us; it may be a deep primeval emotional hope or fear; it may be based on a natural skepticism or intellectual curiosity. But even those of us who believe there is more still accept that reality is revealed to us via our senses, that what we see and hear and feel and taste and smell is real. Some of the most prominent scientists today say that one of the most important lessons emerging from today's scientific inquiry is that human sensual experience is not an adequate guide to the true nature of reality, or at least that reality is far more than is revealed to us via our senses. In fact, if you believe the findings of a large proportion of twenty-first century science—and this is serious science, not fantasy—reality, whatever it is, may be inherently unknowable.

Brian Greene, one of the most prominent and most popular of a number of qualified physicists who write about the phenomenal realities they uncover, says that "the overarching lesson that has emerged from scientific inquiry over the last century is that human experience is often a misleading guide to the true nature of reality. Lying just beneath the surface of the everyday is a world we'd hardly recognize."[1]

Science writer K. C. Cole begins his book, *Mind over Matter*, by quoting British physicist James Jeans:

> We each live our mental life in a prison-house from which there is no escape. It is our body; and its only communication with the outer world is through our sense organs. These form windows through which we can look on to the outer world and acquire knowledge of it. However, these windows are cloudy, veiled by expectations, distorted by frames of reference, disturbed by our very attempts to look.[2]

Now, of course, this insight is not new. Philosophers and theologians, and indeed scientists, throughout the ages have argued that the true nature of reality can never be ascertained by our senses, that it is way beyond sense experience, that reality, like truth, may never be ascertained at all.

Some have taken the position that what we perceive

through our senses is illusion; others that it is real but only a small part of reality.

For much of human history, religious dogma of one kind or another has provided truths and thus certainty. Great theologians and philosophers have always maintained that these truths were based essentially on faith and were not accessible to reason or the senses alone. But people being people and organizations being organizations, the need for intellectual certainty prevailed.

When the old religious truths were undermined by the rise of science during the Enlightenment of the seventeenth and eighteenth centuries, the need many people had for certainty coalesced into scientific truths. From that time to the present, the dogma and myth of the truths of science has built a wall of certainty around the process and the results of science in the minds of the general public and of many scientists.

What is intellectually exciting for me is that the certainties and truths of science have been undermined by science itself. This is the only way science could have been challenged successfully in our modern secular society.

A new understanding of science comes from two revolutionary scientific theories of the last century—relativity and quantum mechanics. These are the two pillars of modern science. Relativity changed the supposed reality of the universe—of space and of time—and quantum mechanics changed the supposed reality of the subatomic world, which in turn changed, and is still changing, the supposed reality of everything. Quantum theory and its various spin-off theories have destabilized all previous scientific truths as well as the commonsense realities we all live by. As one of the founders of quantum theory, the Danish physicist Niels Bohr is reported to have stated: "Anyone who is not shocked by quantum theory has not understood it."

The first quantum theory was formulated by the German physicist Max Planck in 1900 when he suggested that light could be delivered only in quantized units (quanta), essentially discrete chunks. Five years later, in 1905, Albert Einstein contributed further to the theory when he established that light quanta were real, "massless particles" and not just mathematical abstractions. Over the next several years, quantum theory was developed further by Niels Bohr; the French physicist Louis de Broglie; the Austrian Erwin Shrodinger; the German-born Max Born; and the elucidator of the famous uncertainty principle, Werner Heisenberg.

In its early years, quantum theory was just too radical for most scientists. They had to make a leap of faith before they could accept its premises, which were so different from the classical theories with which they were familiar and which formed the truths of science. Some of the early pioneers were never converted to the new truths of quantum mechanics nor were they able to accept its inherent uncertainties. Most notable among the unconverted was Albert Einstein, who expressed his deep-seated reservation with the famous remark during one of his debates with Niels Bohr that "God does not play dice with the universe."

However, most scientists did eventually accept the truth, or rather truths, of quantum mechanics. Today, both for theory and practice, quantum theory dominates the field. Much of modern science has evolved from it; statistical mechanics, particle physics, chemistry, cosmology (the history of the cosmos), molecular biology, evolutionary biology, and geology were all either invented or revised as a result of its development. The practical conveniences of our modern digital world—such as computers, DVD players, and digital cameras—would not be possible without the modern electronics that developed from quantum theory.[3]

"The only 'failure' of quantum theory is its inability to provide a natural framework for our prejudices about the workings of the universe (Wojciech H. Aurek). It is often stated that of all the theories proposed in [the twentieth] century, the silliest is quantum theory. Some say that the only thing that quantum theory has going for it, in fact, is that it is unquestionably correct (Richard Fehnman)."[4]

"Einstein said that if quantum mechanics is right, then the world is crazy. Well, Einstein was right. The world is crazy (Daniel Greenberger)."[5]

It is the truly extraordinary, unbelievable, mind-boggling, but scientifically sound theories about reality that have their basis in quantum mechanics and are now beginning to be communicated by competent physicists and science writers, that have the potential to change the way that even the general public views the world. The old certainties of science have been undermined. They are mostly all gone, as Hans Reichbach says: "Gone is the ideal of a universe whose course follows strict rules, a predetermined cosmos that unwinds itself like an unwinding clock. Gone is the ideal of the scientist who knows the absolute truth."[6]

The late physicist Frank Oppenheimer used to get angry when people would tell him to behave or to believe in a certain way because that was the "real world." "Wrong," he would say. "It's not the real world. It's a world we made up."[7] On the basis of the discoveries of modern science, it is clear that the world around us is made by our perceptions. Certainly, a major preoccupation of science is trying to show the difference between the real world and the "world we made up." But quantum theory has moved the scientific yardstick further and further away from anything that could sensibly be called real. We know, for example, that the real things around us, furniture, houses, trees, mountains, et cetera, are all made of molecules, which are made up of atoms. How real is an atom? As it turns out, not very.

While quantum reality has been known for nearly a hundred years, it is still a comparatively new reality, certainly for nonphysicists. In science—from at least the time of Isaac Newton in the late seventeenth and early eighteenth centuries up to and including Albert Einstein in the first half of the twentieth century to most everybody even today who thinks about and perhaps comments on the world and reality—there are at least three truths that are taken for granted. One is called "realism" and is the belief that the physical world around us actually consists of "real things," of objects that exist independently of us seeing them. Another is called "locality," which is the assumption that something in one place can affect something in another place only if there is enough time for a signal of some kind to travel between the two places and that this signal cannot travel faster than the speed of light. This was especially relevant to Einstein, as his theories of relativity fixed the speed of light as an absolute reality and a base for all other measurements and assumptions. The third is "determinism," the belief that every present or future event is an effect of past causes.

It does appear that quantum theory undermines all three of these truths. Indeed, according to one, if not the major, school of quantum theory—the Copenhagen school named after the location of its founder, Niels Bohr—reality is far more subjective than objective. Just as all experimentation and observation of the quantum world is conditioned by the subject doing the experiments and the looking, this theory holds that reality is created by human consciousness and that there is no objective reality outside of the human mind, or, if there is, it isn't what we experience as reality.

As I will explore, this is very close to, if not identical to, the philosophy of idealism. This philosophical perspective has historically competed with another, the philosophy of materialism or realism. The truths of classical science in the West have been distinctly material. This remains the case for much of science. However, it does appear that the new truths connect again to idealism.

Because of the new insights provided by quantum theory and its derivative theories, extraordinary speculations are being made by respected scientists about the nature of reality. Many examples can be found in the book *Warped Passages* by Lisa Randall.

> The universe has its secrets. Extra dimensions of space might be one of them.... Physical laws—not to mention common sense—have bolstered the belief in three dimensions, quelling any suspicion that there might be more. But spacetime (the combination of the three dimensions of space and the one dimension of time) could be dramatically different from anything you've ever imagined. No physical theory we know of dictates that there should be only three dimensions of space.... Just as "up-down" is a different direction from "left-right" or "forward-backward," (the three dimensions we experience) other completely new dimensions could exist in our cosmos.... Research into extra dimensions has also led to other remarkable concepts—ones that might fulfill a science fiction aficionado's fantasy—such as parallel universes, warped geometry, and three-dimensional sinkholes.[8]

> We could be living in a three-dimensional pocket of space, even though the rest of the universe behaves as if it is higher-dimensional. This result opens a host of new possibilities for the fabric of spacetime, which could consist of distinct regions, each appearing to contain a different number of dimensions. Not only are we not the center of the universe, as Copernicus shocked the world by suggesting five hundred years ago, but we just might be living in an isolated neighborhood with three spatial dimensions that's part of a higher-dimensional cosmos.[9]

Humans have always tended to pattern their domestic, social, and political arrangements after the dominant vision of physical reality, the dominant vision of truth. In the Middle Ages, when virtually everyone believed the world to be the personal creation of a divine being, society mirrored the hierarchy that supposedly existed in the heavens. Dante pictured this world as a series of concentric spheres—heaven, then the planets, down through Earth to the seven circles of hell. This gave everything and everybody a proper place from the divine right of kings down to the lowliest peasant. The prevailing worldview in the Middle Ages was marked by the assurance that man was the all-important, even controlling, fact of the universe. Nature was thought to exist for man and to be immediately present and intelligible to his mind. The whole universe was a small, finite place, and it was man's place.

The Newtonian scientific revolution toppled this worldview and replaced it with a physics of ordinary matter governed by mathematical laws rather than divine command.

> Just as it was thoroughly natural for medieval thinkers to view nature as subservient to man's knowledge, purpose and destiny, so now it has become natural to view her as existing and operating in her own self-contained independence, and so far as man's ultimate relation to her is clear at all, to consider this knowledge and purpose somehow produced by her, and his destiny wholly dependent on her.[10]

This classical view of the world is the commonsense, secular view of the mechanistic, materialistic world that we still live in. However, just as Newton shattered the medieval view, first Einstein's theory of relativity and then the modern quantum theory have irreparably smashed Newton's mechanistic view of the world. Indeed, the Copenhagen and other similar interpretations of quantum theory put man at the center of the universe again. What now appears to be a scientific certainty is that the world is not a deterministic mechanism, as Newton and science since his time believed. What the world actually is today cannot be ascertained with the same certainty.[11]

The subatomic, quantum world—the world of the infinitely small—is an uncertain world. All of the certainties that were the basis of the classical world of Newton and even the world of Einstein have been undermined by theories arising from the study and understanding of this world. One cannot know the exact position and the exact velocity of any subatomic particle. When you know one you cannot know the other. Any kind of measurement encounters great difficulty. The best one can do in quantum theory is to predict the probability that this or that may happen. This is the famous uncertainty principle elucidated by one of the founders of quantum theory, Werner Heisenberg.

The quantum world appears to be a surging world of energy and of potential for the realization of matter (i.e., for the evolution out of pure energy of little things that make up the constituents of matter we see around us). Since it is a world of energy, one may even describe this as a sort of "spirit-world." You can make the best measurement possible about how things are today, but you can only predict the probability of how they will be in the future or that things were this way or that in the past.

In addition—and this is what really set "the cat among the pigeons" in the scientific community and still is the cause of disagreements—the experimenter conditions the results of the experiment. This was a revolutionary thought for experimental science, which always valued its objectivity in treating what was being observed as an "object" separate from the person doing the observing. This separation was the basis of experimental science. But the founders of quantum mechanics saw that, in this subatomic world at least, there is a fundamental link between the observer and the thing being observed.

Until a particle is actually observed, for example, it can be located anywhere and have any speed. It may even be in many states at the same time. When a measurement is made, however, the measurement forces the particle to collapse into a single state. While some states are more likely than others—hence the accuracy of probability—no certainty exists. The universe, according to quantum mechanics, participates in a game of chance. We tend to realize it (i.e., the universe), perhaps even create it, by observing it.

"This is a most contentious issue in quantum mechanics. It is called the 'measurement problem' and is generally regarded as the most important and profound issue posed by quantum mechanics."[12] This perspective is fundamentally different from the traditional scientific as well as the commonsense perspectives. From its beginning as natural philosophy with the Greeks even until today for many scientists who study particle physics, matter is made up of "little things." For the Greeks and for everybody up to practically the twentieth century, these "little things" were called atoms. For today's physicists they are called elementary particles, such as the electron, but include a whole array of additional subatomic particles. However, according to most interpretations of quantum mechanics, the basic events of the physical world are more like energy than matter. They have been described as events that flash in and out of existence.[13] The locations of their appearances are subject to an inherent randomness.

A German radical philosopher of the mid-eighteenth century, Ludwig von Feuerbach, beautifully summarized the revolution in thought brought about by the philosophy of historical materialism introduced by Newton: "The old world made spirit parent of matter. The new makes matter parent of spirit."[14] The new revolution in thought brought about by quantum mechanics reverses this and once again makes spirit parent of matter.

How deep this revolution will penetrate in the public mind remains to be seen. As with the Newtonian revolution, it will require the intellectual understanding and involvement of many competent artists and writers in order to communicate these new truths to the general public. It is worth a caution, however, to wonder what this new worldview, if it is ever communicated generally, would do to the mind of the West. The West has been built on certainty, both scientific and religious in their own domains. Can a philosophy of uncertainty be viable for a dynamic and progressive society?

While many scientists are adept at dealing with uncertainty and the scientific method is based on questioning all certainties, science qua science, just like religion, is a search for certainty, for reality and truth. Modern science, in the minds of most people, is still a world of certainty. Even though science has a limited purview and truly great scientists understand that so-called truths in science are really only approximations or theories, it is considered by many people, including scientists, to be not only the most reliable but the only avenue to truth. This is based on its method of measurement, quantification, and verification through repeated experimentation. To repeat the quote from Nobel Laureate physicist Robert Laughlin cited in the introduction:

> The existence of universal quantities that can be measured with certainty is the anchor of physical science.... Deep inside every physical scientist is the belief that measurement accuracy is the only fail-safe means of distinguishing what is true from what one imagines and even of defining what true means.[15]

While this may be a perfectly legitimate position for a physical scientist to take, so long as it is restricted to physical science, it is a very limiting position when extended to the whole of reality. It assumes an objective world, outside and independent of the observer. It assumes a commonsense view of the world. It assumes that matter, as defined by physical science, is all there is. But when it is thus assumed, it is no longer a scientific proposition; it has become a metaphysical one.

But now these metaphysical truths have been undermined—the old commonsense certainties: of matter that is not dependent on human observation or interaction, of laboratory experimentation that can be independently validated, of measurement accuracy, of the whole basis of reality, even if there is any reality that we can know. Nick Herbert claims:

> No development of modern science has had a more profound impact on human thinking than the advent of quantum theory. Wrenched out of centuries-old thought patterns, physicists of a generation ago found themselves compelled to embrace a new metaphysic. The distress that this reorientation caused continues to the present day. Basically physicists have suffered a severe loss: their hold on reality.[16]

If this is true for physicists, should it not be doubly true for the rest of us? Is this a bad thing?

I believe that it is not and that we should agree with and be comforted by Einstein's comment, published in an article in 1931, that "the most beautiful experience we can have is the mysterious. It is the fundamental emotion which stands at the cradle of true art and true science."[17] Einstein was referring to feelings of awe and exhilaration that we experience when we are faced with things we do not know, particularly when we know we do not know. His message is that one can be exhilarated, not frightened, when one stands on the edge between what is known and unknown. The questioning and pursuit of truth has its own rewards. And, of course, one can—indeed must—pursue truth in the faith that it does exist and is worth pursuing.

Einstein, and arguably all great scientists, had such a faith. His faith may have been in the "clockmaker," the "old one," and/or "the laws of the universe," but it was a strong faith. Others, with equally strong though perhaps quite different faiths than Einstein (indeed even traditional religious faiths), should feel equally exhilarated by thinking about, exploring, and examining the mysterious. The new theories, derived from quantum mechanics, are the most mysterious—and perhaps the most exhilarating and exiting—theories at play today, metaphysically, philosophically, and theologically.

Chapter Two
The Evolution of Theoretical Physics: Newton to Einstein to Bohr

There is some debate as to exactly when the modern age of science began, but it is reasonable to peg its beginning to the work of Copernicus, Galileo, Descartes, and most of all, Isaac Newton (1642–1727).

The most important scientist among those just mentioned, and the one whose work ushered in the "age of science" and ended the "age of religion," was Newton. With some wonderful and creative mathematical equations, he was able to synthesize what was known about motion on earth and in the heavens into the truth of gravity. In doing so, he established what has come to be known as classical physics. Newton clearly explained the concepts of space and time, declaring them absolute and immutable entities, totally separate from each other—truths from the mind of God that showed how the universe was run. All motion was relative to the fixed space and time. The universe was likened to the ticking of a huge clock: having been wound up and set in motion by God, the clockwork universe ticks away under precise rules from moment to moment with complete regularity.

Robert Laughlin explains that Newton changed history by elucidating the scientific case for universal physical law. Newton went beyond the simple observation of regularity in the natural world to identify mathematical relationships that were simple, could always be applied, and simultaneously accounted for what appeared to be unrelated behaviors. His laws of motion were so trustworthy that incompatibility with them became a reliable indicator of false assumptions. The great influence of Newton's treatise, outlined in his classic work *The Principia: Mathematical Principles of Natural Philosophy*, came not from its explanation of planetary orbits and tides but from its use of these things to demonstrate the legitimacy of the clockwork universe—the concept that things tomorrow, the day after, and forever are completely determined from things happening now through a set of simple rules and nothing else.[1]

In his book, *The Metaphysical Foundations of Modern Science*, E. A. Burtt emphasizes the extraordinary influence that Newton had not only on the scientific community but also on the general intellectual mind-set of the West. He profoundly influenced the thinking of the average, intelligent person. During the eighteenth century, he was adored all over Europe. His scientific discoveries of the laws of motion and the law of universal gravitation represented a unique and important victory of the mind. The term *enlightenment*, which was adopted as the title for this period of European history, reflects, more than anything else, the influence of Newton. The veneration shown to Newton was captured beautifully by the poet and writer Alexander Pope in his famous couplet: "Nature and Nature's laws lay hid in night; God said, 'Let Newton be,' and all was light."[2]

Burtt summarizes Newton's contributions as the further development of calculus; the first clear statement of the union of experimental and mathematical methods, which became the norm in science; in giving vague terms such as *force* and *mass* precise quantitative meanings that were amenable to mathematical treatment; and in giving new meaning to space, time, and motion. Burtt concludes by stating that

> in his treatment of such ultimate concepts, together with his doctrine of primary and secondary qualities, his notion of the nature of the physical universe and of its relation to human knowledge, in a word his decisive portrayal of ultimate postulates of the new science and its successful method as they appeared to him, Newton was constituting himself a philosopher rather than a scientist as we now distinguish them. He was presenting a metaphysical groundwork for the mathematical march of mind.[3]

For Newton, science was composed of laws stating the mathematical behavior

> of nature solely—laws clearly deducible from phenomena and exactly verifiable in phenomena—everything further is to be swept out of science which thus becomes a body of absolutely certain truth about the doings of the physical world. By his intimate union of the mathematical and the experimental methods, Newton believed himself to have indissolubly allied the ideal exactitude of the one with the constant empirical reference of the other. Science is the exact mathematical formulation of the processes of the natural world.[4]

This mathematical march, along with the march of human reason to uncover truth, has continued to the present day. It has provided the metaphysical infrastructure of our modern secular society—relegating God and man to distinctly secondary positions. God, when considered at all, is but the original clockmaker. Man is a puny and local spectator, an irrelevant product of an infinite self-moving engine, which existed eternally before him and will exist eternally after him. Mathematical relationships are enshrined. Ideal imaginations are banished into impotence. The scientific metaphysic of Newton ascribes reality to the world of mathematics, which world is identified with material bodies moving in space and time. This metaphysic has conditioned practically the whole of modern Western exact thinking.

While his findings were theories, approximations,

and metaphors, they were considered truths by the intellectual leaders of Europe and then by the general public. Every age and culture has its own metaphors for the realities and truths of the universe. It is this ultimate picture that an age forms of the nature of the world that provides the controlling factor in our thinking. These Newtonian truths lasted until the twentieth century and still monopolize the common mind.

Newtonian, or classical, physics is remarkably accurate for most measurements: gravitational force, speed, distance, mass—pretty much anything in the world around us. The only areas where Newton's physics is inadequate are those areas—the infinitely large and the infinitely small—encompassed by relativity and quantum mechanics.

In the spring of 1905, Albert Einstein came to the conclusion that space and time are not independent and absolute but interconnected beyond common experience. That year he produced four major papers that changed science forever. He won the Nobel Prize for one, describing how light could behave like both a wave and a particle—a duality that established the foundation for the next revolution in science, quantum mechanics. The second paper confirmed the existence of molecules and atoms.

The third was revolutionary. Similar to much of Einstein's theorizing, his third paper that year was based on a thought experiment: if one could travel at the speed of light, what would a light wave look like, and would one perceive time and space differently? His conclusion became known as the second theory of relativity, which affirmed that no matter how fast one is moving toward or away from a source of light, the speed of that light beam is the same—a constant 186,282 miles per second. The speed of light is always observed as constant, regardless of the speed of the observer. He reasoned that if the speed of light is constant then space and time must be relative. Theoretically, if one were on a train moving at this speed, time on the train would slow down, and the train would appear shorter and heavier for someone not moving with it. If one were on a spaceship traveling at the speed of light, when one returned to earth one would find that those who had remained on earth had aged more than oneself had.

His fourth paper showed that energy and matter are different manifestations of the same thing. Their relationship is described by what has become the most famous equation in physics: $e=mc^2$ (energy equals mass multiplied by the speed of light squared). Einstein's theory of special relativity declares that the combined speed of any object's motion through space and its motion through time is always precisely equal to the speed of light. The two motions are always complementary. The maximum speed through space is reached when all light-speed motion through time is fully diverted into light-speed motion through space. When this happens, time stops. If one were to travel at this speed—and this can happen only in science fiction or in theoretical physics—one's watch or clock would stop, and one would stop aging. This is why, in science fiction, if a person travels through space at light speeds, or even at just extraordinarily fast speeds, he or she ages more slowly than those people left at home.[5]

These papers were incomplete because he did not include acceleration or gravity. However, they were complemented in 1916 with the publication of Einstein's general theory of relativity. This theory was also based on a thought experiment. If one were to fall freely through space, one would not feel one's weight. Anyone can experience a bit of this when riding an elevator that begins a fast descent. One momentarily feels lighter. Or when the elevator moves quickly upward, one feels momentarily heavier. This discovery led to a link between acceleration, falling freely, and the concept of gravity—the force that gives one the sensation of weight. Einstein concluded that acceleration under the influence of gravity and freefalling were indistinguishable. Only those observers who feel no force at all—including the force of gravity—can be certain that they are not accelerating. After many years of thinking about this and the mathematical equations that would come to describe how objects move in response to gravity, Einstein came up with his general theory.

Brian Greene provides a different, but complementary, explanation of Einstein's insight, explaining that since gravity and acceleration are equivalent, if one feels gravity's influence one must be accelerating. He uses the example of a person, Barney, who jumps from his upper-story window. Normally, one would describe him as accelerating down toward the surface of the earth. However, Einstein would say that Barney is not accelerating, that he feels no force, that he is weightless, and that this provides the standard against which all motion should be compared. In fact, when one is calmly relaxing at home, one is accelerating. From Barney's perspective as he falls by someone's window, that person and the earth and all other things are accelerating upward. In the classic, perhaps apocryphal, story of Newton's discovery of gravity by having an apple fall from a tree and hit him on the head, Einstein would argue that it was Newton's head that rushed up to meet the apple.

Now this is a very different way of thinking about motion, but it is anchored in the recognition that one feels gravity's influence only when one resists it. By contrast, when one fully gives in to gravity, one doesn't feel it. Assuming that one, and Barney, is not subject to any other influences, such as air resistance, when one gives in to gravity and allows oneself to fall freely, one feels as one would if one were floating in empty space—a perspective that we consider to be unaccelerated.[6]

Newtonian physics explains gravity as a force that acts over a distance. The gravitational force law summarizes how gravity depends on mass and distance, saying that the force of gravity between two masses is proportional to the mass of each—the more massive the objects, the greater the gravitational attraction. The law also explains how the gravitational force depends on the distance between the two objects, as it is equal to the inverse square of their separation.

General relativity presented a dramatic revision of the classical concept of gravity. Einstein explained gravity as a warping or curving of space itself, under the influence of matter. Einstein believed that space is a fabric and is not empty but filled with "stuff." Matter distorts the fabric of the cosmos, and we experience this as gravity. Various pictorial analogies have been used to visualize this force. One uses the example of a flat sheet of rubber that one rolls a marble across. The marble rolls along in a straight line. But, if you imagine a heavy ball resting at the center of the rubber sheet, you can imagine that it will warp or distort everything in its vicinity. Thus, when the marble is rolled across the rubber, its path will curve because of the way the rubber sheet warps around the ball. Einstein showed that it is this warping—because of large pieces of matter such as the sun and the earth and the moon—that we experience as gravity. It is via this warping that the sun holds the earth in its orbit and the earth holds the moon in its gravitational grip.

According to Einstein's theory, matter warps time as well as space. For example, a clock in a strong gravitational field will run more slowly that the same clock would in a weak field. This means that a clock at the base of a mountain will run more slowly that an identical clock at the peak of the mountain, because the lower clock is more affected by earth's gravity.

The conclusion he drew from all of this was that space and time are not separate and distinct entities but an interconnected entity: spacetime. Gravity is a warping of spacetime. Space becomes curved. Planets move around the sun because the sun has warped the space around it, not because of some sort of gravitational pull. Just as his earlier theories paved the way to an understanding of the subatomic quantum world, Einstein's general theory of relativity established the basis for an understanding of the very large: the universe, its foundation in the so-called Big Bang, and mysterious black holes.

Lisa Randall explains that the reason Einstein's conclusion seems so surprising and strange is that our upbringing here on Earth, with a planet that feels stationary beneath our feet, biases our intuition.

> When the force of the Earth keeps you stationary on the ground, you notice the effects of gravity because you are not following the path toward the center of the Earth that gravity would follow. On Earth, we're accustomed to gravity making things fall. But falling really means falling relative to us. If we were falling along with a dropped ball, as we would be in a free-falling elevator, the ball would not go down any faster than we would. We therefore would not see it drop.... In a review paper he wrote in 1907 about relativity, Einstein explains how the gravitational field has only a relative existence, "because for an observer falling freely from the roof of a house there exists—at least in his immediate surroundings—no gravitational field."[7]

The Newtonian system was based on common sense.

A second on earth was the same as a second anywhere in the universe. A foot distance or a pound weight is likewise the same in any location. Because Einstein understood that if the speed of light is constant no matter how fast or slow one moves, then time must get slower the faster one moves and lengths must contract and masses must increase as one approaches the speed of light. Thus space and time become relative. This overthrew the Newtonian universe of classical physics.

With his theory of general relativity, Einstein toppled another of Newton's truths. According to Newton, gravity traveled instantly through the universe. But Einstein showed that nothing travels faster than light. As a result, Einstein theorized that space and time are not empty but instead are like a fabric that can curve and stretch.

> A further effect of this revolution—a momentous one—is to make space and time, as aspects of space-time, active participants in physical interaction.... "Matter tells spacetime how to curve and spacetime tells matter how to move." Gone, therefore, is the conception of space as merely a passive arena for the play of events; and gone too the conception of time as merely the road along which the march of history proceeds. In general relativity, these give way to an actively involved space-time—one that "feels" the presence of matter and energy and "kicks back" accordingly.[8]

So special and general relativity undercut the clockwork metaphor. They showed that there is no single, preferred, universal clock. There is no agreement on what constitutes a moment or a "now." These theories challenged the gravitational theories of Newton—his mathematical framework of how gravity worked. But they did not undermine Newton's theories of predictability and determinism.

As indicated earlier, classical physics is deterministic; the past and the future are rigidly determined. In theory, if one could know the positions and velocities of all objects and particles at a particular moment, one could tell their positions and velocities at any other moment, past or future. Einstein accepted this general deterministic notion of the reality of the universe. Even with his revolution in thinking and his relativity theories about the universe undercutting much of what Newton had said, he accepted the premise that if by some means one could know the state of the universe at any particular moment—where every single elementary particle is, and how fast its going (obviously an impossible task even before the revolutionary uncertainty principle)—then you could predict everything about the future of the universe and explain everything about its past.

As referenced, Einstein won the Nobel Prize for his 1905 paper that described how light could behave like both a wave and a particle. Light was known to flow in continuous waves, but he showed that it could also move along in discrete particles or quanta. He therefore discovered the photon and became one of the founders of quantum mechanics. But he refused to go further with this theory. The wave-particle duality was unsettling.

When the Danish physicist Niels Bohr showed that the electrons in atoms must also behave as quanta, Einstein became very troubled. He realized, in thinking about the quantum interactions between matter and light, that neither the timing nor the direction of the photons spontaneously emitted from atoms could be calculated. The emissions were random. Chance seemed to be the principle element of the subatomic realm. The best one could do was predict probabilities. Einstein refused to accept the reality of this. He believed that a physicist who could not predict individual events was no physicist at all and that this kind of physics was just guesswork. This realm in no way corresponded to the reality that Einstein believed in. "That [God] would choose to play dice with the world," Einstein wrote, "is something that I cannot believe for a single moment." Bohr supposedly replied, "Stop telling God what to do."[9]

It was at a conference in 1927 that Einstein clashed with Bohr. The subject for debate at the conference was "What does quantum mechanics have to say about reality?" This remains a debatable topic. Is the quantum world reality? Is the commonsense, classical world that we experience reality? Is there an objective reality? Or is reality subjective and a function of our consciousness? These questions were first raised during the debate between Einstein and Bohr in 1927.

Einstein was a realist who firmly believed that there is an objective world, an objective reality—separate and distinct from what humans observe. Bohr championed the strange new insights being introduced by quantum mechanics, one of which was that any single elementary particle, whether an electron, a proton, or a photon, does not occupy any definite position until someone locates it. He argued that until one actually observes a particle, one can never know where it is. It has no concrete position but exists only as a probability.

Einstein could not accept the new uncertainty of quantum theory, nor could he accept that not only does God play dice with the universe but our observations influence the roll of the dice. "Do you really believe the moon is not there unless we are looking at it?" he asked. He believed that quantum theory was flawed and incomplete and that a fuller theory—one that would incorporate both quantum mechanics and relativity—would emerge.

> In a letter to Max Born, dated 4 December 1926, he writes: "Quantum mechanics is certainly imposing. But an inner voice tells me that it is not yet the real thing. The theory has a lot, but does not really bring us any closer to the secret of the Old One. I, at any rate, am convinced that He is not playing dice." Einstein favoured a view whereby the only genuine collapse, in a measurement, is a collapse from ignorance to knowledge.[10]

The debates between Bohr and Einstein, indeed their feud, went on until Einstein's death in 1955.

In fact, each had a very different understanding of the role of science. According to Einstein, the role of science is to describe nature as it is. According to Bohr, the role of science is to describe what we can know about nature, which includes the results of all conceivable perceptions and experiments. Einstein believed that one could study nature objectively, that one could separate the experiment from the experimenter and find out about reality as it really is. Bohr believed that one can never separate the experiment from the experimenter, that just as there is no reality without one's perception of reality, there are no truths of nature outside one's perception and understanding of truths. "Bohr said that 'there is no quantum world, there is only an abstract quantum description.' … For him, atoms are not 'little things,' they did not have attributes of their own. Rather, atoms demonstrated certain attributes in different experiments, but those attributes belonged to the setups as much as they belonged to the atoms."[11]

As indicated earlier, another fundamental truth of classical physics and relativity—locality—also appears to be undermined by quantum theory. Like realism, locality appears to be a self-evident truth. When two objects have a space between them, we consider them to be separate and distinct things. Space is the medium that separates and distinguishes one object from another. Things located in different areas of space are different things. In order for one thing to influence another, it has to negotiate the distance that separates them. And, as noted, the fastest that one thing can influence another is bounded by the speed of light. This is common sense and follows the rules of relativity as established by Einstein. Both common experience and repeatable experiments confirm locality. However, there are strong indications in quantum theory and in some experiments that something we do over here can be subtly connected to something that happens over there, without sending anything from here to there.[12] This sounds very weird, indeed. Einstein called it spooky.

But this non-locality appears to conform to the laws of quantum mechanics. These laws indicate that the universe includes interconnections that are not local. Something that happens here can instantaneously be connected to something that happens over there, even if nothing travels from here to there. Subatomic particles appear to have these kinds of interconnections or entanglements, and there can be quantum connections between two particles even if they are on opposite sides of the universe. Even though they are trillions of miles apart, it's as if they are right on top of each other.[13]

While there are disagreements in scientific circles as to what this connection actually is, if one particle truly connects to another, then the connection is instantaneous—faster than the speed of light. Certainly, if quantum entanglement is what it appears to be, space is not what it appears to be. Nor, apparently, is time. "For we convinced physicists, the distinction between past, present, and future is only an illusion, however persistent. The only thing that's real is the whole of spacetime."[14]

So our conventional notions of both space and time are not true in the quantum world. Rather than a distinct separation between objects in space and events in time, objects in space and events in time appear to be inextricably connected, or entangled, in the quantum world. Nothing we experience or hold true in our classical world remains true in the subatomic world. At least two important questions emerge from this recognition. The first relates to the truth of one or the other worlds. Is the reality of the quantum world the true reality? Is our world an illusion, real only in the sense that we make it real by being in it? Or are both worlds real, and is the discovery of the quantum world in the last century simply an expansion of our understanding of the real?

Bohr won his debate with Einstein and was finally vindicated by experimentation in 1982. But by 1935, Einstein believed that he had refuted quantum mechanics. For the next twenty years he worked to find a unified field theory that would combine quantum mechanics and relativity. He never found it, but scientists all over the world are still searching for it today.

Some, perhaps many, respected physicists today are returning to Einstein's belief that quantum theory is incomplete; in some measure, at least, this is because of the bizarre nature of the reality it describes. Lee Smolin explains that

> quantum theory contains within it some apparent conceptual paradoxes that even after eighty years remain unsolved. An electron appears to be both a wave and a particle. So does light. Moreover, the theory gives only statistical predictions of subatomic behavior.... The theory yields only probabilities. A particle—an atomic electron, say—can be anywhere until we measure its state. All this suggests that quantum theory does not tell the whole story.[15]

Yet other physicists, such as Julian Barbour, say that the continuing refusal of many physicists to accept the bizarre realities of quantum mechanics is simply a reflection of their biases, of their fundamental faith in realism.

> I am not persuaded that the people who ought to be best at this—theoretical physicists—do achieve full freedom of thought. Many are passionately committed to an objectively existing external world. They hate anything that smacks of solipsism (the theory that nothing exists or is real but the self) or creationism. This explains the controversies, virulent at time, about the reality of atoms that took place a century ago, and the equally impassioned debates today about the meaning of quantum mechanics (in many ways a continuation of the debate about atoms). For scientists committed

> to realism, atoms that remain the same in themselves and merely move in space and time are very welcome. Atoms, space, and time are the things that either existed for ever or else came into being with the Big Bang.[16]

Barbour questions all of these and other truths in his book, *The End of Time*.

At the heart of quantum theory is the question of whether an atom or an electron or any elementary particle, is a "little thing," or just an abstract construct—energy, spirit—influenced fundamentally by being observed. If an elementary particle exists as an independent entity, then it should have a location and a definite motion. But quantum theory and Heisenberg's uncertainty principle deny this.

They say that you can't know both where an elementary particle is located and how it is moving at one and the same time. They say that a particle is also a wave, and a wave a particle. Measurement appears to determine which. Position and motion form two mutually incompatible aspects of reality in the subatomic world. According to Bohr and much of present quantum theory, the fuzzy and uncertain world of the atom only sharpens into concrete reality when an observation is made. In the absence of an observation, the atom is a ghost.

So, what are the implications of all of this? Perhaps, so long as this bizarre reality is restricted to the subatomic world, it may not matter to many people that the concrete reality that we all see "out there" is not really real. After all, a tree is still a tree, and a table still is a table, right? Well, probably not. Trees and tables, like ourselves, are made of atoms. How can ghosts come together to make something real and solid? And what about the role of the observer in all of this?

"The quantum theory demolishes some cherished commonsense concepts about the nature of reality. By blurring the distinction between subject and object, cause and effect, it introduces a strong holistic element into our worldview."[17] A central paradox of the quantum theory is the unique role played by the mind in determining reality. The act of observation seems to cause the ghostlike features of the quantum world to change into concrete reality. Apparently, an elementary particle cannot make a choice. It has to be observed before a particular outcome is realized.

Lisa Randall explains that quantum mechanics is difficult to understand because its consequences are both counterintuitive and surprising. Its principles are radically different from those underlying all previously known theories in physics, and they also go counter to our own experiences. She says that one reason that quantum mechanics seems so bizarre is that we are not physiologically equipped to perceive the quantum world. Quantum effects generally become significant only at distances of about the size of an atom, so we cannot see what is happening in the quantum world without special equipment—in the tiniest areas of elementary particles, we cannot see at all. Another reason why quantum mechanics is so difficult to understand and explain is that while it is sufficiently far-reaching to incorporate classical Newtonian predictions, there is no range of size for which classical mechanics will generate quantum predictions. Therefore, we cannot understand quantum mechanics by using familiar classical terms and concepts.[18]

The quantum world—the subatomic world of electrons, quarks, et al.—is mysterious even to the physicists working with it. To try to deal with it, to understand it even a bit, at the least many of our assumptions and certainties about reality have to be questioned, if not thrown out. It is an uncertain, chaotic world, a world of pulsating *energy* rather than, or at least as well as, *matter*. In this world, it appears that energy turns into matter and matter into energy, in accordance with Einstein's equation, $e=mc^2$. The quantum world is an amorphous world that can only be understood in terms of probabilities and that is, in large measure, beyond laboratory experimentation and open only to mathematical theory. Yet this world spins off all kinds of practical benefits, including all our modern digital technologies, such as computers. It is a real world, not an imaginary one, even though the theories emerging from the increasing understanding of it are beyond most people's imaginations.

My understanding is that this quantum world is fundamentally a world of energy and that, because of this, it may also be described, at least in a generic sense, as a world of spirit. This is not a term that would be used by physicists. Indeed, it would be anathema to most of them. But the fact is that if, at its most basic, everything is energy, then it is not much of a leap—of faith or otherwise—from energy to spirit.

Even the more traditional particle physicists speak of the smallest objects in the micro world as "massless" objects. This description may work for their mathematical equations, but it is a contradiction in terms. Massless objects are not objects. Those physicists who have embraced the latest string theories, which we will discuss in the upcoming chapter, describe these most elementary of particles as "strands of energy." To my mind, none of these are "little things." At this most elementary, subatomic level, all is energy. This is not incompatible with the term *spirit*, except for the fact that, in the minds of many, *spirit* connotes "mind" and "purpose." This is probably the major reason why physicists would reject it. But while I appreciate the danger in using the term, I will use it, since I am not a physicist.

Chapter Three

The Various "Realities" of Quantum Theory

In his book, *Quantum Reality*, Nick Herbert summarizes the views of what he calls the eight realities of quantum theory.[1] As fantastic as they may seem—and they are fantastic—these are serious scientific theories. His summary provides a "shorthand" appreciation of the range of thinking and just how incredible the theories really are.

1) The Copenhagen interpretation, Part I: **There is no deep reality**. This is the position taken by Niels Bohr. He believed that the world of the senses—our world—floats on a subatomic world that is not real. He said, "There is no quantum world. There is only an abstract quantum description." While this point of view provides great latitude for interpretations about the quantum world (including those that use terms such as *spirit*)—interpretations that see this world as essentially an uncertain world of energy that somehow morphs into matter to create our world—it is not a crank position within the physics community. In fact, the view that

"there is no deep reality" represents a point of view of a majority of the physics community. It is called the Copenhagen interpretation because it was developed at Bohr's Copenhagen institution. This interpretation of the quantum world explains why so many outstanding and knowledgeable physicists have stated that nobody really understands quantum theory or the quantum world.

2) The Copenhagen interpretation, Part II: **Reality is created by observation**. This interpretation consists of two distinct parts. The first part says that there is no quantum reality in the absence of observation. This was stated clearly by Bohr and accepted by the majority of physicists. In the quantum world you cannot separate the observation from the observer. The act of observation affects what is observed. The second part is more problematic and is not supported by the majority of the physicists who support the first part. It states that observation creates reality. What we see is real, but these phenomena are not really there in the absence of an observation.

3) **Reality is an undivided wholeness.** This is the quantum interpretation that appears to be similar to the insights of religious mystics and Eastern religions, such as Hinduism, Buddhism, and Taoism. This interpretation takes the position that, despite appearances, the world is actually a seamless and inseparable whole. Reality is what the mystics describe as an "undifferentiated unity." There is no real differentiation between what has been described as "objective outside reality" and

"self-conscious reality"—in other words, between the objective and the subjective. Quantum mechanics clearly shows the inseparable connectedness of the whole universe. That is the fundamental reality.

4) **Reality consists of a steadily increasing number of parallel universes.** This is the many-worlds interpretation—sometimes called "the multiverse" or "parallel universes" interpretation—and it is one of the most bizarre contentions of theoretical physics. It claims that an infinite number of universes are created on the occasion of each act of measurement. This means that for any situation where several different outcomes are possible—literally every decision that every person makes—some physicists believe that all outcomes actually occur. In order for this to happen, entire new universes are created, identical in every detail except for the single outcome that created them. Since each person makes these kinds of decisions all the time, there are billions upon billions of universes—literally an infinite number since they are continually being created.

5) **The world obeys a nonhuman kind of reasoning.** This interpretation says that a new kind of quantum logic is required to cope with the new quantum reality. Physicists who hold this view argue that the quantum revolution goes so deep that our concepts and reasoning do not suffice. Quantum theorist David Finkelstein summarizes this position by saying,

> Einstein threw out the classical concept of time;
> Bohr throws out the classical concept of truth.

> Our classical ideas of logic are simply wrong in a basic practical way. The next step is to learn to think in right ways, to learn to think quantum-logically.

6) **The world is made of ordinary objects.** This is the realistic argument. It says that an ordinary object possesses attributes of its own, whether it is observed or not (i.e., an object exists "from its own side"). The commonsense simplicity of this argument has convinced a few physicists that it can be extended into the quantum realm. They believe, for example, that atoms are "little things." This is heresy to many theoretical physicists today. Although the realist position was held by Einstein and was a basis for the famous Einstein/Bohr debates, the majority of physicists sided with Bohr. As we have seen, some eminent physicists still do take a realist position.

7) **Consciousness creates reality.** While this appears similar to quantum reality number three, it differs in that it accepts that entities—that is, entities such as quons at the quantum level—in the physical world are real in that they exist independently of mind, but their dynamic attributes, including momentum and position, are mind-created. These dynamic attributes build the molecular structures that create the physical things of the world. Thus, the reality we see around us, which is made up of these subatomic particles, is created by the human mind. This may seem unduly complicated, but the theory is supported by outstanding physicists

and mathematicians. Consciousness-created reality adherents include light/matter physicist Walter Heitler, Nobel laureate Eugene Wigner, and mathematician John von Neumann.

Many of the extraordinary quantum issues discussed in books by Herbert and others were first outlined in Neumann's book on quantum theory. The so-called *von Neumann proof* states that if quantum theory is correct—and it has been proven correct many times—then the world cannot be made of ordinary objects. This means that the realist or neorealist interpretation is logically impossible. Physical objects would have no attributes if a conscious observer were not watching them. While the basic ingredients may be there, independent of mind, at the quantum level, their position and momentum are mind-created. Thus, the reality around us, the world that we see and experience, is created by the mind.

8) **The world is twofold, consisting of potentials and actualities.** Many physicists accept the Copenhagen interpretation, which states that there is no deep reality (Part I) and that observation creates reality (Part II). Both of these assert that only phenomena are real and that the world beneath—the quantum world—is not. However, this begs the question of how exactly the reality we see and experience comes about. Out of what is this reality created? Since the nature of unmeasured reality is unobservable, many physicists dismiss such questions. However, since it describes measured reality with exactness, quantum theory may contain some

clues to the raw material that provides the basis of the phenomena we experience.

One of the clearest descriptions of the uncertainty surrounding the experimental processes involving the subatomic realm is given by Shimon Malin in his book, *Nature Loves to Hide*. In fact, this process, and the uncertainty that governs it, is central to his major thesis: that the subjective is reality, not the objective. Malin takes the view, like Bohr and the Copenhagen school to which he obviously belongs, that there are no "little things" in the subatomic world. Rather, it is a world of pulsating energy, a world of uncertainty, but also a world of potentiality. An actual measurement—of the position and/or movement of an electron—conducted by a scientist in a laboratory is what converts the element in this world from potential to actual.

Illustrating the uncertainty principle, Malin explains that one can use a TV screen to measure the position of an electron. When an electron hits the screen, there is a flash of light, which illustrates the position of the electron at that instant. However, there is no way to predict exactly where the electron will hit the screen. All the experimenter has is a probability distribution, the knowledge of how probable it is for the electron to arrive at different points. Some locations are more probable than others, but there is no way to predict the place of impingement. Before impinging on the screen, the electron simply does not exist in space; rather, it is a field of potentialities. However, when it impinges on the screen, it does actually exist. This Malin calls an elementary quantum event, or an event in space and time. Through its interaction with the TV screen, which is a measuring apparatus designed to determine an electron's position, the electron transitions from potential to actual existence. This transition is called "a collapse of the quantum state (or wave function) of the electron." Malin explains that the word *collapse* refers to the fact that before the transition, the whole screen is available for the electron; it could hit anywhere. However, once the electron becomes actual, it is found somewhere on the screen. The possibilities collapsed to a single location, and there is no probability of finding it anywhere else.[2]

Some logical questions in all of this are: Why does a collapse occur? When does it occur? What brings it about? Malin quotes another of the founders of quantum mechanics, Paul Dirac, as answering that "nature makes a choice." By his use of the term *nature* one can assume that he refers to the natural world around us, but at the quantum level. This is a truly fascinating perspective because not only does it indicate that reality comes out of potentiality during observation, in this case an experiment, but it also demonstrates that the reality that exhibits is somehow decided by nature itself.[3] It claims that nature makes and acts on a choice. The choice is an act of creation; the transition from the potential to actual creates an actual event. And, so far, no scientist has yet discovered or explained the mechanism for the collapse or how—let alone why—nature chooses one possibility over another. Because of this, Malin is adamant that the scientific principle of objectivation—treating nature as an object—has to be transcended, because he believes that the universe and its constituents are alive. The stuff of the universe is not matter, at least not as we ordinarily think of it, but experiences and potentialities for experiences.[4] In this way, it can be spoken of as spirit. This puts human beings at the center of the universe.

This perspective, indeed the whole Copenhagen interpretation, reflects the idealism of one of history's greatest philosophers: Plato. His worldview is wonderfully summarized by the Allegory of the Cave in the seventh book of *The Republic*. This allegory presents a vision of reality consisting of three major levels of being. The highest and most real is "the Good," which is the source of beings of the Intelligible realm of the many Forms or Ideas (other than the Good), the next level down. The last, and lowest, level is the sensible world of transient phenomena in space and time. These are mere shadows of the Forms—shadows because they have no independent existence. The source of their existence is the being of the Forms. Humans who believe in the reality of the senses mistakenly believe that the sensible world exists independently and is the only reality there is. According to the allegory, humans are like prisoners in a cave who mistake shadows for the objects that cast the shadows.

Since Plato is the father of Western philosophy, variations of this idealistic worldview are fairly common. For example, in the third century AD, the Neoplatonist sage Plotinus suggested a multilevel concept of reality:

One, Nous, Soul, nature, and the sensible world. The One—which is a counterpart to Plato's Good—is the supreme principle of unity and is beyond thought. The Nous, on the other hand, is the principle of being. The other levels descend from there.

Another good example of an idealistic worldview is given by the English philosopher George Berkeley, who proposed his view of idealism in the eighteenth century as a philosophical antidote to the truths of Newtonian determinism. He said that material objects do not exist until they are perceived by the conscious mind, or in a deeper sense, in the mind of God. I will look at this possible connection, as well as the general idealist position, more fully later. Berkeley's position is interesting on its own accord, but also because Berkeley was ridiculed during his lifetime and has not been taken as seriously as he probably should be by students of philosophy. During his life, his theory was famously refuted by the English essayist Samuel Johnson, who kicked a rock and said, "I refute it thus." Now we know, of course, that the stone is made of atoms, which are mostly empty space.

For Plato, Plotinus, and Berkeley, the sensible world is just a reflection of the principle of being, however it is described. For science—at least for classical science—the sensible world is the only reality there is. However, Malin explains that the Nous is not entirely absent from science. The idea that the world of the senses is sustained by an invisible substratum is a fundamental premise of science. This premise is called "the laws of nature."[5] The laws are unifying principles that regulate events in the physical world. The difference between these and the philosophy of idealism is that science considers the laws lifeless. This is a necessary consequence of the principle of objectivation. The removal of oneself, the observer, from the domain of nature that one is trying to understand is the removal of life from nature. Quantum theory insists that the observer cannot be divorced from what is being observed in the subatomic world. Thus, objectivation, which has governed science since the time of Newton, has been undermined.

However far one wants to speculate regarding quantum reality—and there appears to be little or no limit to such speculation even among highly qualified theoretical physicists—the logic of the whole subject area is truly weird. Apparently, everything that happens in our world arises out of possibilities in the quantum world. But, also apparently, this quantum world is no more substantial than a promise. Not only can physicists not agree on a single theory of reality in the quantum world, they are not even sure that a correct theory is on their existing list. In the meantime they—and we—exist within the reality available to our senses, common sense, or classical reality.

Let me again emphasize that these findings, the many quite extraordinary theories of quantum mechanics are not the products of science fiction but have been developed and are being developed, debated, and in some cases, debunked by the best scientific minds of the modern world. Whether or not the moon would be there if nobody were to look at it is a debatable point, but at least it is debatable. There is a strong theoretical possibility that the moon would not be there if nobody were to observe it—in fact, that there would be no thing, nothing, no universe, no world if nobody were to observe them; that it is human consciousness that creates whatever reality we experience. As has been suggested, perhaps as pre-Copernicans believed, man is indeed at the center of the universe.

Regardless of what theoretical perspective a nonscientist reader is able to accept, the one obvious certainty is the extraordinary openness for creative intellectual and even spiritual exploration presented by the various quantum theories. Everything we see and experience—reality as we see and understand it, the world, the universe, everything—is, at its basic subatomic level, pure energy. Out of this energy—sometime, somehow—particles or "little things" develop, and this goes on to develop the universe, or our world and us.

How this happens is being studied experimentally right now, in 2009, by international physicists using a new billion-dollar, high-energy particle accelerator, which is the highest-energy accelerator ever built and is buried underground near Geneva. The theory physicists using the accelerator are attempting to validate is that a mysterious element known as the Higgs field, or the Higgs boson, somehow imbues energy with mass. This is why the boson has been popularly dubbed the "God particle."

While scientists say they are trying to show *why* particles have mass, it seems to me that if this experiment

Chapter Four:

Strings and Branes and the Illusion of Space and Time

So essentially, at the end of the twentieth century, we ended up with three secular truths or versions of truth—or metaphors, if you will—for how the universe is conceived. Classical physics used the clockwork metaphor to describe the universe. Special and general relativity introduced some subtleties into the clockwork metaphor. As Brian Greene explains, "there is no single, preferred, universal clock; there is no consensus on what constitutes a moment, what constitutes a *now*. But the universe unfolds with the same regularity and predictability as in the Newtonian framework."[1] Quantum mechanics broke with both of these traditions. Its metaphor is uncertainty and probability, if that can accurately be called a metaphor. Each truth or metaphor appears to apply to a particular realm: the first to our everyday world; the second to the world of the very large, like stars and galaxies; the third to the world of the very small, like atoms and electrons.

For decades, physicists have been aware that the weird

and uncertain quantum world does not mesh, practically or theoretically, with either the classical world that we experience or the principles of relativity that apply to the very large, to the universe. A central challenge in theoretical physics is how to mesh quantum theory and relativity into one unified theory, a so-called theory of everything. It is hoped, indeed expected by many theoretical physicists, that string theory, in one or other of its theoretical manifestations, will do this.

The Standard Model of particle physics takes the position that, at the smallest micro-atomic level, there are still particles. They are envisaged as formless and massless points, whatever that means, but they are particles nonetheless. Particle physics has been, and still is, a bulwark of theoretical physics. It has helped provide an understanding of phenomena, from elementary particles to the universe. It answers questions about forces in the universe: electromagnetic force, the weak force associated with nuclear decay, and the strong force that binds together quarks with protons and neutrons. However, there are plenty of mysteries that this theory does not provide answers for, including the nature of dark matter and dark energy that provide the vast majority of the "stuff" of the universe, and how gravity fits into an overall picture of the universe. String theory, or really its evolution as superstring or M-theory, is the leading candidate for a unified theory that will provide answers to these questions and unify theories of the very big (the universe) with the very small (the quantum world).

Brian Greene has been a popular elucidator of this theory. He explains that "superstring theory starts off by proposing a new answer to an old question: what are the smallest, indivisible constituents of matter."[2] The ancient Greeks called these atoms. Only relatively recently, in the latter part of the nineteenth century, was it confirmed that there is indeed something called an atom. However, further research showed that it is not an indivisible constituent of matter and that it can be divided into smaller and smaller units. Atoms consist of electrons that swarm around a central nucleus that is, itself, composed of even smaller constituents of matter: protons and neutrons. These, in turn, are composed of even smaller constituents; quarks, that, in Greene's words, "can be modeled as dots that are indivisible and that have no size and no internal structure."[3]

This is clearly a mathematical hypothesis, since a piece of matter "of no size" makes no sense—not that common sense matters much when trying to understand quantum theories. The mathematical hypotheses of some scientists allow for even smaller particles than quarks, and many of them ad infinitum. As Robert Laughlin explains in *A Different Universe*, "The natural world is thus an interdependent hierarchy of descent not unlike Jonathan Swift's society of fleas:

> So naturalists observe, the flea
> Has smaller fleas that on him prey;
> And these have smaller still to bite'em
> And so proceed ad infinitum."[4]

String theory gets rid of all the infinitely numbered and infinitely small particles and substitutes strings, or filaments of energy, which vibrate this way and that, and in doing so, the vibrations produce elementary particles. So, the basic ingredient is not a point, a dot of no size, but rather a string that has no thickness, only length, and thus a one-dimensional entity.

However, when one gets to this level physicists speculate that these are not particles as conventional thinking supposes a particle to be, but rather they are forms of energy that combine to form particles that do have size, however tiny, and can be measured. After all, the sizes are unthinkably small. Greene says that the strings are some hundred billion billion times smaller than a single atom's nucleus, which itself is much smaller than a single atom.[5] Another description says that a string would be smaller than a virus in the same proportion that a marble is smaller than the entire universe; or, a string is in proportion to an atom as an atom is to the solar system; or, it would take a trillion trillion strings to cross the smallest atom; or, enlarging a superstring to the size of a real string is comparable to making a virus bigger than the entire Milky Way galaxy.[6] Clearly, these are not "things" in any traditional understanding of that term.

The fathers of quantum theory did not believe that elementary particles are "little things;" rather that their fullest description is given not in words but in mathematical symbols and equations. Heisenberg concluded that "the smallest units of matter are, in fact, not physical objects in the ordinary sense of the word; they are forms, structures, or—in Plato's sense—Ideas, which can be unambiguously spoken of only in the language of mathematics."[7]

It is really the "beauty of the math, of the equations," that is the basis of string theory, which may be best described as mathematical metaphysics. There is no experimental evidence to support it. Nobody has ever seen a string, nor will anyone ever see one. What the math and the theory confirm is that infinitely small strings vibrate like a violin string to produce particles such as protons, neutrons, and a whole variety of atoms and molecules that are the building blocks of our world. "The mass of a particle in string theory is nothing but the energy of its vibrating string."[8] This sounds like pure energy. Brian Greene sounds like he supports this point of view when he describes strings as "tiny filaments of energy."[9]

This theory of strings and superstrings certainly does open up the field to wide philosophical and even mystical speculation. But then, so does quantum theory in general, even if many in the physics community are appalled by this elaboration. As an example, this description is closer to the description of the infinitely small in the book *The Tao of Physics* by physicist Fritjof Capra, who says that all is energy, that there are no indivisible constituents of matter.

Both Greene and Capra use Einstein's equation to prove their points. It shows that mass and energy are interchangeable. Energy is produced from mass, and mass is produced from energy. Greene explains that the mass of a particle in string theory is simply the energy of its vibrating string. Capra and others do not use the string explanation but explain that, ultimately, the very smallest of the smallest particles are produced from energy.

Because there is no evidence that strings actually exist nor any immediate potential to assess their existence, string theory does have its critics. However, the theory, in various forms, has many highly respected proponents. Indeed, strings are the most popular undiscovered objects in the history of theoretical physics.[10] Lisa Randall explains that, in many physics departments in the 1980s, superstring theory superseded particle physics, and the so-called superstring revolution was like a coup. Many physicists went so far as to think of it as the supreme theory that underlies everything, the "TOE," the "Theory of Everything."[11]

A main point of criticism of string and superstring theory is the impossibility of seeing strings. But I'm not sure why this should be such a problem, since you also can't see the points with no dimension that are the key elements of the Standard Model of particle physics. In any event, science—physics in particular—has always worked within a theoretical framework of things that one cannot see: the atom being but one example.

From its very theoretical beginning, the premise of the atom was established and pursued because it was a logical premise and because it implied the existence of things that could be seen. This is the same for the particle physics theory of subatomic matter—electrons, protons, quarks, et cetera. Some of these can be seen, or at least their reality can be observed, but they were projected in theory before they could be observed because they provided a consistent mathematical logic to the way the world is.

This is what the current theories are doing, including superstring theory. Even though strings cannot be seen and very likely never will be seen, for many of the most prominent theoretical physicists, they provide the best theoretical basis for an understanding of reality.

As well as the problem of size and the resulting impossibility of ever seeing a string, or probably ever demonstrating experimentally that one exists, there is a further objection to superstring theory. To make the mathematical theories of superstrings work, space must have more than three dimensions; as many as nine or ten or eleven. As we know, the three-dimensional universe available to our senses and or common sense is forward/back, side to side, up/down—latitude, longitude, and altitude. Einstein added a fourth dimension: time. While impossible for our common sense to visualize, and putting aside their science fiction potential, this multidimensional theoretical construct is taken very seriously by theoretical physicists. They believe that these hidden dimensions contain clues about the nature of gravity, the origin of the universe, and the identity of that most mysterious entity, dark energy/matter.

After string theory was developed in the 1970s and 1980s, there evolved five different versions, all requiring nine space dimensions. However, in 1995 the leading string theorist, Edward Witten, realized that they were all just different variations of one underlying theory. Witten named this underlying, superstring theory *M-theory*. He said it could stand for magic, mystery, or membrane. It was membrane that tended to stick, because it introduced a new dimension to string theory. Instead of nine dimensions in the old theories, M-theory requires ten; eleven if time is included. Rather than all objects being one-dimensional strings, higher-dimensional objects called membranes were permitted as well.

The theory of membranes—now called branes—has fairly recently become a hot topic in science. M-theory combines the concept of strings with that of membranes. The two are considered interchangeable. Understanding how one can be like the other is based on a complex concept in physics called duality. This refers to a special kind of symmetry in which two different descriptions of something turn out to be equivalent. A good example is the electron, which sometimes acts like a wave and sometimes acts like a particle.

While the extra dimensions were originally thought of as being very small and curled up, by the end of the last century, new theories suggested that they might be very big and not curled up but rather very thin like a membrane and spread out, perhaps over huge distances. These theories claim that it might be the reality of branes, perhaps very many of them, that accounts for why gravity is so much weaker than the other forces. The theory is that if gravity is the only force that propagates in the extra dimensions, this could account for why it is comparatively weak in our three-dimensional world.

Even more extraordinary, while our universe appears to be huge in the familiar three dimensions of space, with additional dimensions the universe might be extremely thin, the way a sheet of paper is big in two dimensions but thin in a third.

Some theories suggest two-dimensional branes that are like a huge drive-in movie: very thin, but as high and as wide as one could imagine. Other theories suggest three-dimensional branes, so if a single brane were very large, perhaps infinitely large, it would fill all of the three dimensions we can sense. In fact, some theorists suggest that our entire cosmos—the whole of spacetime of which we are aware—may be a huge brane. We may be living in a braneworld. There could be many more such braneworlds right next to ours but completely invisible to us. If we visualize a brane as a membrane, like a thin piece of paper, but with three dimensions, then our universe could be just one piece, with many pieces on either side of ours. Our universe could be much like one page in a very large book.

There are several variations of this hypothesis, but all are of this type: extraordinary and unbelievable in the extreme. With this visual image—as probably inaccurate as it may be, given that it is a visual image, and by definition, we cannot visualize extra dimensions—we can understand why gravity might be comparatively weak if it is the only force moving through all the braneworlds.

Once physicists began to speculate about big branes and our entire universe being a brane, they began to speculate as to why branes might be invisible. Since ordinary matter, like light, can't get to where they are, it is impossible for us to see other branes, even though they may only be the thickness of a sheet of paper away from us. In this new view of the very large and very thin brane, all ordinary forms of matter and energy stick to the surface of the three-dimensional, or so-called three-brane, universe. So all things—light, radio waves, magnetism, electrons, all sorts of fundamental particles—could be stuck on the brane, and all operate in our familiar three-dimensions of space. The sole exception is gravity. Gravity, and only gravity, is allowed to move into the extra-dimensional space.

As could be expected with this theoretical construct, imagination, even the imaginations of theoretical physicists, can run wild. With extra dimensions, there are a huge number of possibilities for the structure of space. Countless universes could fit into the extra dimensions. Such universes, or braneworlds, might contain all kinds of different and exotic forms of matter, possibly with stars, planets, and people. Perhaps we see only three dimensions because we are stuck on this particular brane, while other surrounding branes have more dimensions. Perhaps matter can move in and out of our four-dimensional spacetime region, appearing and disappearing as it enters and leaves. Perhaps the so-called dark matter and dark energy, which have never been seen but make up some 96 percent of the universe, are stuck on a brane, or branes. Any of this would be very hard to prove experimentally, but it all fits into a theoretical framework that has many physicists very excited.

Now, of course, we may not be able to see or otherwise detect these branes and the additional dimensions because they are not there. They may simply be effects of various mathematical and theoretical scenarios gone wild. However, all of this is based on a great deal of rigorous thought and analysis by a great many very competent people. This does not guarantee that it is correct, but it does guarantee the seriousness of the theories and the process. Current research is trying to develop a new understanding of cosmology that incorporates these new insights of superstring or M-theory.

Certainly, speculative theoretical physics has entered a remarkable era. Ideas that at one time would be described as science fiction are now routinely being considered at least theoretically, and possibly even experimentally, possible. New theoretical discoveries about strings and extra dimensions have radically changed how physicists, astrophysicists, and cosmologists think about the world and reality. It is very probable that, as the saying goes, "We ain't seen nothin' yet."

For example, near the conclusion of *Warped Passages*, Lisa Randall asks us to reflect on what we know for certain about the dimensions of space. Not very much, as it turns out. We don't know that space everywhere is three-dimensional; we know only that space that we can see looks three-dimensional. Space beyond our ability to see might go into inaccessible territory. After all, she claims, the speed of light is finite, and our universe has existed for only a finite amount of time, which means that we can only know the surrounding space within the distance that light could have traveled since the beginning of the universe. While this is a very long way, it is not an infinitely long way. The dividing line—if you will—between this distance and whatever may be beyond it is known as the horizon. This is the dividing line between information that is and is not accessible to us. Beyond this horizon, we don't know anything. Space need not look like ours.[12]

This is very similar speculation to the question of what existed before the Big Bang, or what existed before God created the universe. We simply do not know, and furthermore, assuming of course that there actually was something like the Big Bang, we are unlikely ever to know, even after finding a "theory of everything." The exact bang, or start, of the Big Bang is what cosmologists call a singularity. At that point, and certainly before it, all mathematical equations and laws of physics break down. We do not and cannot know anything.

Physicist Michael Lockwood explains that, strictly speaking, we should regard this singularity as the limit of spacetime rather than a part of it. One is therefore not supposed to ask what happened before the Big Bang, since there was no before. St. Augustine dealt with the same question in the fourth century AD: "What was God doing before He created Heaven and Earth?" asked the heretical sect of the Manicheans. The trite answer was "Preparing Hell for people who ask irreverent questions."

But St. Augustine took the question very seriously and believed that time itself came into being only at the Creation. Because of that, there is no before in relation to God's creation of Heaven and Earth, and God himself must be conceived as being outside of time.[13]

Paul Davies, a physicist and cosmologist as well as a brilliant writer of accessible and popular books on these sorts of metaphysical questions, agrees that the most serious objection to the causal version of this cosmological argument is that both *cause* and *effect* are concepts embedded in the concept of time. The idea that there was no before, because the Big Bang represented the beginning of time itself is a standard answer of science. However, he believes that this begs a question. If time really did have a beginning, any attempt to explain it in terms of causes must be done in a much broader and deeper conception of cause than we know. He says that many physicists have suggested that space and time are not fundamental concepts but approximations. Just as matter is made up of atoms, and beyond them of elementary particles, so might spacetime be built out of more primitive, more abstract entities.[14]

One of the important lessons of the many extraordinary discoveries of the last ten years could very well be that "space and time may be doomed." One leading string theorist asserted "I am almost certain that space and time are illusions," while another imagined that "space and time could turn out to be emergent properties of a very different-looking theory."[15]

Chapter Five

Our Mysterious Universe: Dark Matter, Dark Energy, and the Big Bang

As we have seen, the infinitely small—the quantum realm of electrons, quarks, strings, et al.—is a realm open to extraordinary speculation. But so is the infinitely large, the realm of the universe. The realities of both the big and the small are turning out to be equally dubious and difficult to comprehend.

Some cosmologists, in studying the past and present of the universe, are increasingly unsure of what they are actually seeing. They are coming to believe that some of the galaxies in space are, in fact, illusory copies of galaxies that have already been identified. This could be the case if the shape of space is not simple, as has been believed, but twisted and complex. This is a subject for mathematicians who specialize in topology, or the shape of space. On very large scales, space's topology could be what is called *nontrivial*—that is, twisted around in such a way that it is possible to see the same galaxy at different places in the sky. More data will be needed to determine if all galactic images are real or if some are ghosts and to determine whether the universe is finite or infinite. If it turns out that the topology of space is nontrivial, it would force an extraordinary restructuring of our conceptions of space.

Perhaps the greatest area for speculation surrounds questions relating to dark matter and dark energy. Whatever substance makes up about 96 percent of the universe is unknown. Not only does this unknown entity form the bulk of all matter/energy, including our realities on earth, but its properties appear to determine why the universe is the way it is. This is a huge mystery, arguably one of the greatest mysteries in science, if not the single greatest, maybe, in anything. Essentially, we live in an environment (our universe) that we know comparatively little about.

The need for such matter to exist in the universe was confirmed in the 1970s by astronomer Vera Rubin. She realized that there simply isn't enough visible matter to provide the necessary gravity to keep the stars in place. By all rights, because stars move around in galaxies so fast, they should spin off into space. But they don't. So she reasoned that there must be something that cannot be seen holding them in place.

This something is still unidentified, but it is called dark matter. The name refers to the fact that nobody knows anything about it rather than to the fact that it is actually dark. In fact, it is transparent. It can't be seen because light goes right through it. Dark matter doesn't emit any light. Nobody knows what it is, but it apparently makes up the majority of material "stuff" in the universe.

The other unknown is an even stranger, nonmaterial "something" called dark energy. Astronomers and cosmologists believe that about 73 percent of the universe is made up of this dark energy, about 23 percent is dark matter, and the other 4 percent—just 4 percent—is matter/energy that we know about.[1]

In *Strange Matters*, Tom Siegfried explains that midway through the 1980s, astronomical observations revealed that the universe is far more complex than previously thought. Galaxies or small clusters of galaxies were not, as had been assumed, simply scattered randomly through the cosmos. Astronomers discovered that, in contrast to a random scattering, the universe appears to be an architectural marvel, a network of bubbles and walls stretching across all of visible space. Rather than being scattered throughout space, galaxies seem to congregate along the surfaces of imaginary spheres, like giant bubbles. Some clusters of galaxies seem to be collected around several bubble-like surfaces that form a sort of very long wall, extending some 100 million light-years across the sky.[2]

This discovery confounded earlier theories of the universe, since the Big Bang theory of the beginning of the universe suggested that matter was scattered uniformly through space, with no big lumps or bubbles. But this new research shows that galaxies did not form at random. They formed in groups, or groups of groups, that are separated by giant bubble-like voids in which few galaxies are found. Theories of the origin of the universe, via the Big Bang, suggested that ordinary matter (primarily protons and neutrons) could not coagulate rapidly enough to result in a galaxy like the one proposed by the earlier theories. Some other form of matter must have been present, and dark matter is a convenient name for matter that nobody has found nor knows anything about.

The majority of people know that the primary theory about the beginning of the universe is called the Big Bang. Not everybody knows that this is an unfortunate name. Whatever happened, there was no bang, since nobody was around to hear it. Big Bang is merely a metaphor for something that happened around 13 or 14 billion years ago, when the universe, for an entirely unknown reason, began to expand very rapidly. An explosion is an apt metaphor because the theory says that something infinitely small, infinitely dense, and infinitely hot—sometimes described as a "sizeless point," or in other words a quantum something—in something like a hundred billionth of a trillionth of a second, doubled in size one hundred times, expanding in every direction up to a million trillion trillion times bigger than it started and kept on expanding. Talk about fantastic metaphors; at least the God of the Old Testament took six days to create the universe.

It is important to keep in mind that this is a theoretical construct, a mathematical theory that has been formulated to give an account of reality as it is now perceived by science, just as the Old Testament account of creation was a theoretical account formulated to give an account of reality as it was then understood. Obviously, nobody knows what actually happened. However, scientific evidence suggests a very hot and dense phase for the young universe, like a fireball of an explosion, hence the Big Bang metaphor. The Big Bang notion comes from relatively recent observations that the universe is expanding and cooling and evidence that this expansion has always been the case. Thus it began as a very tightly packed "something."

Out of this creative event came both space and time. Because of this, as we have seen, for many physicists, before the Big Bang has no meaning, since both space and time were created by that event. That need not be the final answer. It is hard to imagine, with new and fantastic theories of both the very small and the very large being proposed almost daily, that this will be the end of the question of reality before the Big Bang. It has always been a classic question in philosophy and theology and now in physics. In fact, the Big Bang theory was based on Einstein's equations of general relativity, and these do not work in the quantum world. This means they do not work at a "sizeless point." So new theories that try to incorporate both quantum theory and relativity are being developed.

Sir Martin Rees says that now that we know the key properties of the universe at the present time, the challenge is to explain how it got that way. While it is possible to trace its history back to about a microsecond after the big bang, nobody knows what happened in that first, formative microsecond. The wide variety of ideas about this—branes, inflation—make it clear that science has a long way to go. Rees says that the easiest theory to understand is eternal inflation, which was proposed by Alan Guth.[3] However, he, like leading theoretical physicists, believes that the basis for real progress will be a theory that relates gravity to the microworld. The necessity of doing this arises because at the very beginning, the entire universe would have been the infinite size of an elementary particle. The universe would have been a quantum world, so there should be an essential link between cosmology and the microworld. String theory and M-theory are the most popular theories to make this link. However, Rees warns that if we cannot link this incomprehensible ten-dimensional theory with anything we can observe, the theory will die.[4]

Alan Guth is the father of the so-called inflationary theory of the universe. His research focuses on the application of elementary particle theory to the early universe. Writing in the book *The New Humanists*, he explains that the inflationary universe theory is an add-on to the standard Big Bang theory. Basically, it adds a description of what drove the universe into expansion in the first place.[5] The classical Big Bang theory was not a theory of the actual bang, but a theory of what happened after it. Even if the universe was infinitely small, like an elementary particle, there was something there to begin with. Inflationary theory—or theories, since there are a number of versions—tries to answer the question of what made the universe "bang."

In the same book, physicist Paul Steinhardt explains further that the data collected about the universe in the last decade have provided what he calls a consensus model and an alternative model. The consensus model involves a combination of the traditional Big Bang model of the 1920s, '30s, and '40s with the inflationary theory proposed by Alan Guth in the 1980s, but with a recent amendment.[6] While the traditional theory and the Guth addition are fascinating, they do not finally account for what came before the bang and why the bang happened in the first place, let alone exactly how everything unfolded as magically and wonderfully as it did to allow for the evolution of human beings. So the amendment, or the alternative model, goes back to a very old belief that says the universe is endless, that it evolves in cycles—birth, death, rebirth.

The consensus model of the big bang theory assumes that time has a beginning. For reasons that nobody is clear about, the universe exploded from nothing into something, full of matter and energy, and developed in a logical way to expand and finally formed earth and life. The alternative model says that there was no beginning. The universe is endless; it is eternal. It evolves cyclically, through periods of hot to cold to hot. We are caught in a middle or moderate period that has allowed life to form, but the universe will eventually cycle into "the big chill," and all life will die. Then the sequence of events will begin again: the universe will warm up, and perhaps some form of life will evolve. We're simply witnessing the latest cycle.

Besides reflecting many ancient myths of some religions, this gets rid of the problem of trying to describe how and when it all started. With an eternal universe, science does not have to explain how it all began or what was before the beginning. This theory of a cyclical, eternal universe is very close to the theological speculations of many Eastern religions.

One significant modification to the Big Bang theory suggests that ours may not be the only universe. There may be multiple, even infinite, universes. Within the cosmological community of scientists, there is a theory called eternal inflation, which refers to a never-ending series of bangs producing new bubble universes. Again, the reason for such theories is to reflect mathematical theories to explain why reality is the way it is. Obviously, nobody can actually see these proposed universes. While science tries to adhere to the concept of experimentally objective findings, scientific theory—especially the new theories—has to deal with issues that cannot be seen or experimentally proven. These universes and other such theoretical constructs may be necessary to fully understand the reality of the universe that we do see. They may be necessary, certainly, to explain the mathematical formulae used to describe reality. Incidentally, this many universes theory is different from the parallel universes theory proposed by Lisa Randall and others, which reflects the braneworlds scenario.

If the universe were proven to be infinite, then it

would include an infinite number of regions of space, equal in size to what astronomers can already see. With no limit on space, all possible atomic arrangements could recur over and over again. As a result, all possible combinations of matter and all sequences of activity could happen somewhere. This might mean that every person and every event would exist in multiple places.[7] Each of us would be infinitely duplicated, and each decision and each event in each of our lives could be equally duplicated with an infinite number of alternative scenarios. Every scenario becomes theoretically possible.

This theory applies to time as well as to space. Some physicists believe there may be a second dimension of time, where the universe evolves in a different way. While the notion may seem extraordinary—and it is—it is no more so than extra space dimensions, and extra space dimensions are a central feature of M-theory. Apparently, two time dimensions are needed to make sense of certain versions of string theory. For reasons that escape the nonmathematician, they are necessary in order to make the math work. And, as with most other current theories, the working of the math is important.

Much of the discussion around the various quantum and cosmological theories sounds a bit like the medieval debate about how many angels could dance on the head of a pin. This medieval topic sounds ridiculous to us today, but it was not at all ridiculous to the medieval philosophers and theologians arguing it. It was based on the nature of a world they believed in—an afterlife of angels and devils, of heaven and hell. It dealt with the nature of corporeality, of the nature of spirit, of the nature of the world, of reality that cannot be seen. It was a discussion, in other words, that fit within the beliefs of the time; just as many of these most fantastic, theories about the nature of the quantum reality and the cosmological reality fit within the beliefs of our time. In my opinion, they are no less strange than the concept of angels dancing on the head of a pin—or on anything else for that matter.

None of this, of course—angels, strings, branes, even massless particles—has ever been observed. The upshot of all of this is many different subtheories and many different imaginary worlds. They may all be mathematically elegant, but to what extent are they real? For centuries the only reality for science was what could be experimentally validated over and over. Still today, many traditional physicists would say that the ultimate hurdle for these and any new theory is experimental testing. This is emphasized by physicist Robert Laughlin. "In physics, correct perceptions differ from mistaken ones in that they get clearer when the experimental accuracy is improved."[8]

It is the lack of such experimental verification for many, even most, of the current popular theories—specifically the various string theories—that is the subject of Lee Smolin's book, *The Trouble with Physics*:

> Only when a theory has been tested and the results agree with the theory do we advance the theory to the rank of true theories. The current

Chapter Six
Mathematical Metaphysics

As much as many physicists would like to maintain the old experimental assurances and hope to see laboratory validation of some of their mathematical theories, the "good old days" are unlikely to return. The quantum world and quantum theories that have been applied to the cosmos and all aspects of reality have moved science way beyond this probability. Current theoretical physicists are treading uncharted metaphysical waters, with only their confidence in the logic of mathematics to support them. But this confidence isn't something new. The guiding principle—the beauty or the elegance of mathematical formulae—that is paramount has always been key to theoretical physics. This elegance is based in the concept of symmetry.

In a recent book, *Why Beauty Is Truth*, mathematician Ian Stewart discusses the evolution of the concept of symmetry, why it is beautiful, and how it revolutionized science and mathematics. He explains that symmetry is a special kind of transformation, or a way to move an object. If the object looks the same after being transformed, then the transformation concerned is symmetrical. For instance, a square looks the same if it is rotated around a right angle. This idea, extended and embellished, is fundamental to science's understanding of the universe and its origins.

> Quantum physics tells us that everything in the universe is built from a collection of very tiny "fundamental" particles. The behavior of these particles is governed by mathematical equations—"laws of nature"—and those laws again possess symmetry. Particles can be transformed mathematically into quite different particles, and these also leave the laws of physics unchanged.... The word "symmetry" has to be reinterpreted as "a symmetry." Objects do not possess symmetry alone; they often possess many different symmetries. What, then, is a symmetry? A symmetry of some mathematical object is a transformation that preserves the object's structure.... The first point to observe is that a symmetry is a process rather than a thing,... a permutation, and permutation is a way to arrange things.... There are three key words in the definition of symmetry: "transformation," "structure," and "preserve."[1]

And, symmetry is beautiful.

While this reliance has been, and is being, criticized

and challenged, one must remember that mathematics has been a singularly successful method of prophecy, or prediscovery, in physics. Math has predicted some very strange things that have later turned out to be experimentally validated.

In physics—if not in other areas of science—somehow humans have been able to discover, via mathematics, basic laws of nature. Mathematical symbols have correctly predicted the existence of a whole range of strange objects and phenomena, such as antimatter, quarks, black holes, radio waves, the expansion of the universe, and the curvature of space. Many scientists believe that the success of mathematics discloses an inherent mathematical structure of the cosmos, which makes it comprehensible. Math's ability to predict the presence of strange "things," or "stuff," in the universe bolsters their argument.

Einstein was convinced that the deep truths of nature are not open to experimentation in the classic sense. He believed that they can be discovered only by "free inventions of the mind," that "the creative principle resides in mathematics," and that "pure thought can grasp reality, as the ancients dreamed."[2] Here, at least, he is very much in agreement with Plato. Paul Davies describes mathematics as "the poetry of logic."[3] He says that the bizarre ideas of quantum mechanics can only be fully grasped by mathematics, because human intuition is an inadequate guide. By using mathematics as language, science can describe situations that are beyond the ability of the human mind to imagine.

While physicists, like everyone else, have mental models of atoms, light waves, the expanding universe, electrons, and so on, these models can be wildly inaccurate and misleading. In fact, Davies says it may be impossible for anyone to correctly visualize certain physical systems, such as atoms, because they contain features that do not exist in the world of our experience.[4]

Pure mathematics is sometimes compared to an art form. Truth is in its beauty and simplicity. Physicists search for beauty and simplicity first and then think about what it might mean. Often, it seems, tuning into this beauty and simplicity provides extraordinary insight into the physical world. Why should this be? Nobody knows for certain, but there are several theories. Perhaps reality is created by our minds. Since math is a product of the mind, it would follow that math would be useful in predicting reality. But if there is something real out there—independent of our minds—then perhaps, since our minds are part of nature, we are naturally tuned in, and this tuning occurs most effectively via mathematics.

The eminent British mathematician Roger Penrose believes that mathematical truths are "out there," like Plato's Ideas or Forms: independent of our minds and available to be discovered by the trained mind.

> How "real" are the objects of the mathematical world? From one point of view it seems that there can be nothing real about them at all.

> Mathematical objects are just concepts.... At the same time there often does appear to be some profound reality about these mathematical concepts, going quite beyond the mental deliberations of any particular mathematician. It is as though human thought is, instead, being guided towards some external truth—a truth which has a reality of its own, and which is revealed only partially to any of us.[5]

Perhaps God is a mathematician, as some of the ancient Greeks supposed. Perhaps the physical world is a giant mathematical system, maybe even like a giant clock, as Newton suggested. According to this perspective, all reality—everything we see and experience, all the elementary particles that we don't see but make up this reality—are just manifestations of the mathematical plan. Whether any of this is the case, what is obvious is that pure mathematics—that is, math of the mind rather than the applied math of technology—has been a most effective prophetic process in uncovering, or discovering, truths about the physical world.

However, some scientists disagree. They argue that mathematics is not a discovery about the external world but rather an invention based on metaphors of the human mind. They argue that mathematical ideas are all metaphors—just as all words are metaphors—that come from the human mind. Through them we impose reality on what we perceive and study. Math is a product of the human brain, not something that is basic to any objective reality. Even if math were to be "out there in the world," there is no way we could possibly know that.

The French mathematician Henri Poincare said that math works because it deals with like things. Since math works for physics, physics must be dealing with like things—not a smorgasbord of disconnected things. Math's success points to a fundamental underlying "likeness," or unity, in the universe, which suggests a connectedness among all things. This reflects the mystical beliefs of both Eastern and Western religions, along with many tenets of Buddhism, Hinduism, and Taoism and the philosophies of Plato, Neoplatonism, and more current philosophies. An undifferentiated unity is a basic understanding of reality in all mystical religious traditions and is being suggested by some of the current quantum theories. However, in all of this, to again quote one of the originators of quantum mechanics, Max Planck: "Great caution must be exercised in using the word *real*." Good advice.

So, nobody quite knows why mathematics has been so successful in predicting things in physics. In its pure form, it has little to do with the real world. The legendary physicist Richard Feynman said that "the glory of mathematics is that we do not have to say what we are talking about." Or as the British mathematician Bertrand Russell famously put it: "Mathematics may be defined as the subject in which we never know what we are talking about, or whether what we are saying is true."[6] While it obviously has been effective at times in predicting real things, it has likely been ineffective at least as many times. However, it is its comparatively logical consistency that makes it an ideal tool in the scientific search for what is real. To emphasize the important point made by Lisa Randall, "A mathematical theory must be internally consistent but, unlike a scientific theory, it has no obligation to correspond to an external physical reality."[7]

Certainly, a nonmathematician can point out that it is esoteric to believe that mathematics provides *the* path to truth and that this path can be known only to those who have been initiated into the sacred mathematical cult. Like the old masters of hieroglyphics, or even the masters of ancient Greek or Latin, the new truths are available only to those who have mastered the esoteric language. Other forms of communication, as well as wars of attrition, put an end to the first, while the rise in literary popularity of vernacular languages, the printing press, and the Reformation ended the second. Perhaps the desire, if not the need, of theoretical physicists to write understandable, even popular, books will undermine this new esotericism. But perhaps not.

There has always been, and likely always will be, a desire to believe that the truth lies in the esoteric. This reflects a universal human desire to believe that we "know what truth is" while others do not. But I'm sure it also reflects the desire of scientists—our modern high priests—not to have to explain their beliefs to the "great unwashed," who might, if pressed, ask uncomfortable and unanswerable questions.

In any event, certain equations, for the professional mathematician, are considered to contain inherent beauty, similar to how great artists and musicians see beauty in their art. The difference, for me, is that while I can appreciate Mozart, I have no appreciation for the equations that reflect string theory or supersymmetry; nor will I ever have. To appreciate the beauty of math, one must become a mathematician, and a very accomplished one. To appreciate great art or great music, one, does not have to become an accomplished musician. One undoubtedly appreciates more if one is an artist or a musician, but being one is not a precondition to a basic appreciation, as it is with math.

Perhaps because I am a nonmathematician, I side with the historian John Lukacs, in his book *At the End of an Age*. He said,

> We have seen that the earlier assumption—that the physical essence of the entire universe would be revealed in our discovery of its original smaller particle—has now degenerated into the second assumption, the myth of the Unified Theory: that many physicists are now inclined to believe that even if we cannot find the smallest building block of the universe, we can find a mathematical formula that will explain the entire universe: a Theory of Everything.... But, most mathematical formulae about atomic matters remain untested and untestable, since they are theoretical and abstract. The belief

> that the universe is written in the language of mathematics is entirely outdated. "What is there exact in mathematics except its own exactitude?" Goethe wrote. He was right, as many mathematicians themselves in the twentieth century have confirmed.[8]

However, physicists respect what mathematics tells them. It seems to have the power of prophecy, and it has been proven over many years, by later laboratory experimentation, to be highly accurate. The reason for its accuracy is its effectiveness in describing the laws of nature. Based on years of experience, physicists have faith in these laws and in something constant beneath the changes in the world. They have faith in the fundamental beauty and symmetry of nature, even while the reason for this beauty and symmetry remains a mystery.

Again, mathematician Ian Stewart emphasizes the mystery and cautions against uncritical faith, implicitly criticizing the vision and faith of Roger Penrose.

> Why does the universe seem so mathematical? Various answers have been proposed, but I find none of them very convincing. The symmetrical relation between mathematical ideas and the physical world, like the symmetry between our senses of beauty and the most profoundly important mathematical forms, is a deep and possibly unsolvable mystery. None of us can

> say why beauty is truth, and truth beauty. We can only contemplate the infinite complexity of the relationship.... Mathematics is not some disembodied version of ultimate truth, as many people used to think.... Mathematics is created by people.... Many beautiful theories have turned out, once confronted with experiments, to be complete nonsense.[9]

A similar caution is given by Lee Smolin:

> One point that string theorists are passionate about is that the theory is beautiful and elegant. This is something of an aesthetic judgment that people may disagree about, so I'm not sure how it should be evaluated. In any case, it has no role in an objective assessment of the accomplishments of the theory.... Lots of beautiful theories have turned out to have nothing to do with nature.[10]

Practical warnings notwithstanding, scientists, like theologians and the rest of us, have a "faith" perspective. They believe in an inherent order of things, in natural mathematical laws. These laws are worth discovering. Einstein is reported to have said that "there is no doubt that all but the crudest scientific work is based on a firm belief, akin to a religious feeling, in the rationality and comprehensibility of the world." While this belief may have been shaken by quantum mechanics and the belief systems that have evolved from current theoretical physics have more questions than answers, science is still based on this faith in the natural order of things, however that order came about.

Ian Stewart is critical of such faith, but admits to having it.

> The view that a Theory of Everything must exist brings to mind monotheistic religion.... So, in many ways fundamental physics is more like fundamentalist physics. Equations on a T-shirt replace an imminent deity.... Despite these reservations, my heart is with the physical fundamentalists. I would like to see a Theory of Everything, and I would be delighted if it were mathematical, beautiful, and true.[11]

Sir Martin Rees references one particular theoretical question that interests some physicists and many philosophers regarding the uniqueness of natural physical laws. This is the so-called fine-tuning argument, which holds that our universe seems to be very special and that it arose because its laws have a very specific character. The slightest deviation, even at the most subatomic level, would mean that the world we know would not exist. Why this world and not some other? Our existence is a mystery, since one can easily imagine laws that would make the world we know impossible.[12] Now, one answer to this question—an answer favored, of course, by theologians—is that the creative mind of God—however that metaphor is imagined—was and is behind the creation of the universe and that is why it is as it is. God is the reason there is "something rather than nothing."

However, Rees, and I'm sure most other scientists, is not prepared to accept this answer. He says that the most natural answer to the mystery of the fine-tuning argument would be that the Big Bang wasn't the only one, that there are many universes. Each is governed by different laws, and only one of which allows structures and life to evolve.[13] Obviously, the many universes theory, examined briefly in chapter 5, is simply a logical means of explaining why there is something rather than nothing, without resorting to a theory beyond mathematics and therefore beyond science.

Science is unable to provide an explanation of why the initial conditions, those at the Big Bang, were as they must have been in order to have resulted in the presently observed universe. The vast majority of initial conditions would have led to a universe, or something, where we would not be around to think about or observe it. Only an exceptional initial condition could have led to the present order. This is a great puzzle for science, and is referred to as the Goldilocks enigma.

> Modern science is in the remarkable position of possessing beautiful and very well tested laws without really being able to explain the universe.... Scientists feel they should not invoke miracles to explain the order we see, but that leaves only statistical arguments, which

> give bleak answers, or the so-called anthropic principle that if the world were not in a highly structured but extremely unlikely state, we should not exist and be here to observe it.[14]

In general, science communicates the truth that the universe is the result of natural processes that can be apprehended and understood mathematically, not requiring any higher power. But since this is clearly a leap of faith, it is no more logical than the leap of faith that says, ultimately, life must be based on something other than pure chance. That something can be described as God, or other names in other cultures. Both of these are faith statements.

Of course, to emphasize, there are many important questions and issues that cannot, by definition, be dealt with by science or mathematics: love, hate, humor, consciousness itself. One could do a scientific or mathematical analysis of a piece of music but never understand why, or even how, it touches the soul. One of the founders of quantum mechanics, Erwin Shrodinger put it succinctly: "Science cannot tell us a word about why music delights us, or why an old song can move us to tears."[15]

Regarding the potential for the development of a theory of everything, which will supposedly combine quantum theory with relativity to produce one master theory of how the universe works, from the infinitely small to the infinitely large, he said that it "would not tell you how life originated, it wouldn't tell us about the nature of consciousness, and it wouldn't tell you how I'm going to vote in the next election."[16] Finally, to quote the formulator of the original quantum theory, Max Planck: "Science cannot solve the ultimate mystery of Nature. And it is because in the last analysis we ourselves are part of the mystery we are trying to solve."[17]

Scientists of all stripes and the theoretical physicists who are the ground of science, are trying to apprehend and understand reality, which, while they might not use this term, is also truth. Sometimes some are adamant that their theory has found truth, or at least is on the right track to finding it. But most understand that it is a process, a pursuit, and that they are unlikely to ever know absolutely what it is. In fact, quite the opposite of what the general public believes, the best scientists know how profoundly little they actually know about reality. They know that their truths are approximations or theories. Their business is to push forward and test the limits of all theories.

Would that metaphysicians of all stripes might feel free to do likewise.

Chapter Seven
The Mystical Tradition

Over the past century, wild and wonderful new theories have evolved in science, specifically in theoretical physics, about space and time, about matter, about reality, and their evolution continues. As well as providing different models of the world around us, these theories have forced physicists, and increasingly the rest of us, to rethink their concepts of reality itself. In some ways these new concepts appear to be closer to the truths and realities of mysticism and of Eastern religions associated with the mystical tradition than those traditionally associated with science.

For most of us, the truth about the world around us, of reality, can be described as naïve realism or naturalism or materialism—the belief that what we see is real and exists "from its own side" even when nobody is there to see it. However, even with this commonsense view, there is room to acknowledge that even though the material world may be real, the world we perceive may not be. Things like solidity and color are hallucinations—

products of the perceiving mind. The universe, in itself, is made up of atoms and elementary particles, and at the most elementary level, they appear to be energy rather than matter at all. However, for practical purposes, "we have agreed that sanity consists in sharing the hallucinations of our neighbors."[1]

If naïve realism/naturalism/materialism is the first great theory about reality, then a second is that of idealism, a common point of view of philosophers throughout the ages. It is currently a popular point of view of quantum theorists and is identified specifically with the Copenhagen school. This theory of reality is far removed from our perceived natural universe with all its "things" and its laws.

While the world of the realist is constructed from observation and the evidence of our senses, the world of the idealist is made from the observation of pure thought. The idealist believes that we are sure of only two things: the existence of our own consciousness (that I am I) and the existence of an object, an idea, with which I deal. In other words, what we know is mind and thought. The universe is really a collection of such thoughts. However, our thinking is limited. We cannot conceive all that is to be conceived. Nor do we combine things in the right order. Reality, for the idealist, can be described as "the big thought;" for some, but by no means all, it is described as "the big thought in the mind of God." We pick up only fragmentary hints of this. Our world of phenomena, which we believe to be real, therefore is made up of mere shadows, like in Plato's Allegory of the Cave.

It is the reality, or truth, of idealism that is believed to be apprehended in mysticism. The next few chapters will examine some aspects of the mystical tradition, of Eastern religion, and of their supposed connections to quantum theory.

In her seminal book *Mysticism*, first published in 1911, Evelyn Underhill beautifully summarizes the goal that unifies the objectives of the mystic, the theologian, and today's theoretical physicists who are pursuing string theory and a unified theory of everything. All are searching for reality and truth; Truth with a capital *T*. Underhill says that

> all men, at one time or another, have fallen in love with the veiled Isis whom they call *Truth*. With most, this has been a passing passion: they have early seen its hopelessness and turned to more practical things. But others remain all their lives the devout lovers of reality: though the manner of their love, the vision which they make to themselves of the beloved object, varies enormously. Some see Truth as Dante saw Beatrice: an adorable yet intangible figure, found in this world yet revealing the next. To others she seems rather an evil but irresistible enchantress: enticing, demanding payment, and betraying her lover at the last. Some have seen her in a test tube, and some in a poet's dream.... Last stage

> of all, the philosophic skeptic has comforted an unsuccessful courtship by assuring himself that his mistress is not really there.[2]

Some writers have gone so far as to suggest that quantum theories provide a surer path to truth (i.e., to God) than does traditional religion. Certainly, theoretical physics has progressed—if progressed is the right word—to a point where questions formerly associated with religion and/or philosophy are being tackled daily by physicists. Many physicists, particularly those who are more traditional, decry this evolution, especially its connections with mysticism and the fact that nonphysicists are becoming interested in all these "mystical" insights. However, quantum theory has opened Pandora's box. Scientists are going to have to get used to the fact that others from other disciplines and interests are going to "get into the game." And the game is philosophical, theological, metaphysical, and mystical speculation.

While all of this speculation coming from quantum theory may be intellectually stimulating for some, it may also be both intellectually and emotionally troubling for others. Remember the ancient Greek myth: when Pandora opened the box that Zeus had given her, she let out human ills into the world, or in a later version, let all human blessings escape and be lost, leaving only hope. In no way do I believe that blessings will be lost with these current metaphysical speculations. What will remain after the questions, other than more questions, will be faith. So if we end with faith and hope, leavened perhaps with love and the knowledge that ultimate questions end with questions and not "provable truths," all will definitely not be lost.

In any event, whatever theory we believe in ultimately reflects a faith position. Whatever metaphysical position we take, we are forced to live and to think and to die in an unknown and probably unknowable world. "The horrors of nihilism, in fact, can only be escaped by the exercise of faith, by a trust in man's innate but strictly irrational instinct for the Real, 'above all reason, beyond all thought,' towards which at its best moments his spirit tends."[3] The fundamental distinction between the faith of science and that of religion and mysticism is the role of human reason; reason's role is primary to the first, secondary but important to the second, and limited in the third.

The great mystics, of East or West, when they report their experiences, say that they have succeeded in establishing an immediate connection between the spirit of man (i.e., themselves) and what they sometimes describe as the only reality, which has been called the Absolute and which some theologians call God. The reality that they describe sounds remarkably similar to that postulated by quantum physics.

Mystics have always distrusted the standard realistic modes of communication and understanding. They have always mistrusted the reality of sense impressions. This intuitive distrust has become the distrust of science. And, even though the mystics also distrust the intellect and science certainly does not, it is reasonable and intellectually exciting to examine the insights of mystics alongside the insights of theoretical physicists. One after another, mystics, and now physicists, have rejected an appeal to the unreal world of appearances, which is the reality of sensible persons. The mystics have always affirmed that there is another way by which one may reach reality:

> More complete in their grasp of experience than the votaries of intellect or sense, they accept as central for life those spiritual messages which are mediated by religion, by beauty, and by pain. More reasonable than the rationalists, they find in the very hunger for reality, which is the mother of all metaphysics, an implicit proof that such reality exists; beyond the ceaseless stream of sensation which besieges consciousness.[4]

Now these similarities, as with all descriptions of reality and truth, should be taken with a grain of salt. We must remain skeptical. But they should not ever be dismissed arbitrarily, especially not on the basis of ignorance of either or both mysticism and quantum physics. We all may be dancing in the dark, but it is a dance in wonder, and we have written records of the best dancers in the world—mystics, philosophers, theologians, and scientists—to choose as dance partners. The key phrase from the quotation above should be central to our understanding, that the hunger for reality is the mother of all metaphysics.

Mysticism, like quantum mechanics or any serious exploration of reality or truth, is a complicated business. *Mystical* and *mysticism*, like *religious experience*, are very broad and general terms. They lack precise definitions. One cannot be sure that two persons are talking about the same thing. There is no mathematical exactness, let alone laboratory experiments that others can reproduce, as there is with science. When the phenomena are carefully defined, as they are by several investigators, the experience is immediately limited, and the investigators often find that they are comparing apples and oranges. The reason, of course, is that, essentially, the experience is beyond the ability of words to express, so any and all verbal expressions are inadequate and can be misleading.

The psychologist William James, in *The Varieties of Religious Experience,* lists four common or universal characteristics of the mystical experience. These are:

1) Ineffability—The subject insists that the experience defies expression, that one cannot adequately report the content.
2) Noetic Quality—The subject feels that the mystical state was one of knowledge, that he gained insight into depths of truth beyond the attainment of his intellect.
3) Transiency—The subject does not remain in the ecstasy of the mystical state for long. Half an hour, or at the most an hour or two, seems to be the limit.
4) Passivity—Although there are ways to facilitate a

> mystical state, such as by fixing the attention or going through certain bodily performances, when the actual mystical state has set in, the subject feels as if his own will were inoperative and sometimes as if he were freed by a higher power.[5]

Evelyn Underhill also gives four rules or notes, in place of the four given by James, that she believed could be applied as tests to determine whether a given case was truly mystical.

1) True mysticism is active and practical, not passive and theoretical. One does not merely have an opinion about it. Rather, it is an organic life process that the whole life does.
2) The mystical experience is wholly transcendental and spiritual. It does not add to, rearrange, or improve anything in the visible universe. Though the mystic does not neglect his duty to the many, his heart is always set upon the changeless One.
3) For the mystic, the One is not merely the Reality of all that is but also a living and personal object of love.
4) The termination of the mystical adventure is living union with this One. This is a form of enhanced life, which is arrived at by an arduous psychological and spiritual process—the so-called Mystic Way.[6]

Mystics tend not to write about their own mystical experience, although some do. They maintain that it is essentially ineffable, unexplainable, an overwhelming personal emotional and absolute experience.

Walter Stace, in his book *The Teachings of the Mystics*,

says that at its base, whether it is expressed by mystics from Eastern lands with religious traditions of Hinduism, Buddhism, or Taoism or by mystics from Western lands with a Judeo-Christian tradition, all fully developed mystical experiences involve the apprehension of what has been described as an ultimate nonsensuous unity in all things, a oneness or a One that is impossible for the reason or the senses to penetrate. The experience totally transcends our sensory-intellectual consciousness; in fact the experience *is* the One.[7]

When this experience gets translated into theological terms, as it does in the formal religious movements, it tends to speak to a pantheistic view of reality, claiming that all is spirit and that all spirit is one. Formal Christianity, in the tradition of Judaism, has always reacted negatively to this view, believing absolutely that God is separate and distinct from his creation. The great Christian mystics who have written about their mystical experiences usually translate it into the accepted orthodoxy. Otherwise, they get excommunicated, or in earlier times, burned at the stake.

However, and more seriously, outstanding commentators on the mystical traditions, such as R. C. Zaehner in *Mysticism, Sacred and Profane*, do not agree that there is this so-called universal core to the mystical experience (i.e., this "undifferentiated unity"). Zaehner makes the point that there are various types of experience, which relate to the cultural and psychological backgrounds of the various mystics. So a Christian mystic, for example, experiences a mystical union with God by love, which

Zaehner would say is a fundamentally different experience than that of a Buddhist or a Hindu mystic. F. C. Happold, in his book *Mysticism*, takes a more compromising position, saying essentially that there may be a similar experience—not necessarily the same but certainly similar—but the experience is always understood and translated according to the culture and understanding of the particular mystic and the mystical tradition.

Another writer on the mystical tradition, Friedrich von Hugel in *The Mystical Element of Religion*, denies that there is a specifically distinct, self-sufficing, purely mystical mode of apprehending reality at all. He says that all the errors of the exclusive mystics are the result of this belief that mysticism does constitute such a separate kind of human experience.

> Mysticism's true, full dignity consists precisely in being, not everything in any one soul, but something in every soul of man; and in presenting, at its fullest, the amplest development, among certain special natures with the help of certain special graces and heroisms, of what, in some degree and form, is present in every truly human soul.[8]

The interpretations of von Hugel, Happold, and Zaehner leave room for a variety of cultural and religious interpretations, get away from the sometimes physically vulgar and—in my mind—purely emotional and more limited aspects of mysticism, and open up the notion to include the wonderful creative insights of people of genius in a whole variety of areas, including science and the arts.

Be that as it may, mysticism and the mystical experience—as it is more exclusively understood as an ecstatic emotional experience brought on by a variety of means—can perhaps best be described as an apprehension of an undifferentiated unity or, as Walter Stace describes it above, as "an ultimate nonsensuous unity," or a Oneness or a One in which the persons experiencing the ecstatic state believe they are participating. The ecstatic state can occur naturally or it may be induced by various means, including sensory deprivation, yoga exercises, and/or drugs.

This apprehension of a One, a unity in all things, is the link that some writers connect to the apprehension of the theoretical physicists that, at its most subatomic level, all matter and reality itself, is energy. This, some writers say, is what the mystics have always experienced: that all is energy, that all is spirit. Humans are energy, spirit, and therefore part of the "all," the One. Whether this connection is justified remains a very open question. But one thing that is not in question is that mystics have always denied that reality is what we experience in our daily lives. This is what many great philosophers have always said and what science is now saying.

William James writes that "our normal consciousness, rational consciousness as we call it, is but one special type of consciousness, whilst all about it, parted from it by the flimsiest of screens, there lie potential forms of consciousness entirely different."[9] It is these forms of consciousness that, it is said, are realized in mystical experiences. These are entirely unlike our everyday consciousness and cannot be described rationally at all. This is why mystics say that their experiences are ineffable.

So while the mystics do describe their experiences as best they can once the experience is over, during the experience itself, there are no conscious thoughts that can be mediated by words. Minus the emotionalism, this is similar to what physicists say about quantum theory. Words are inadequate. Mathematics is the only way to describe this world adequately. The result of this is similar to mystical consciousness: an inability to adequately communicate the experience.

The origin of the word *mystic* is in the Greek *mystikos*, meaning "of mysteries." A mystic was someone who was initiated into these mysteries and therefore gained esoteric knowledge of divine things, in effect having been "reborn into eternity." It was unlawful for the mystic to speak of the secret wisdom thus obtained. In the course of time, certainly by the second or third centuries AD, the meaning of *mystic* had evolved into a more general approach to the problem of reality, in which the intellect, but especially intuition, played a more important role.

As Christianity included more Greek concepts in its early years, specifically those common to Neoplatonism, mystical consciousness became increasingly important. While the perceived basic experience of mysticism, that of an undifferentiated unity, does not appear to be a religious experience, once it became integrated into Western Christianity, the experience took on a religious perspective as it did in the East in relation to Hinduism, Buddhism, and Taoism. In fact, probably anyone who already has a religious perspective would interpret the experience within that perspective and within the culture in which they were born and live.

Stace distinguishes two major types of mystical experience, which he calls extrovertive and introvertive. Both apprehend the One but reach it in different ways. As the names suggest, the extrovertive way looks outward through the physical senses to find the One in the external world. This is not a bad description of the mystic way of quantum physicists. The introvertive way is to turn inward and find the One at the bottom of the self. This latter way predominates in the mystical tradition and, even in theology, finds expression in some of the greatest minds in the history of the Catholic Church, notably St. Augustine. Both quantum physicists and St. Augustine would decry any such connections with the mystical tradition, relying instead on intellect and faith. But quantum physicists are looking outward and finding what sound very much like mystical truths in the realities that they find, while St. Augustine and other theologians and philosophers have extolled the virtues of looking inward to find God and faith at the basis of one's being.

F. C. Happold takes a strong view that

> mysticism has its fount in what is the raw material of all religion and is also the inspiration of much of philosophy, poetry, art, and music, a consciousness of a *beyond*, of something which, though it is interwoven with it, is not of the external world of material phenomena, of an *unseen* over and above the seen. In the true mystic there is an extension of normal consciousness, a release of latent powers, and a widening of vision, so that aspects of truth unplumbed by the rational intellect are revealed to him.[10]

Again, this feeling of a "beyond" can be either within or without.

Classical mysticism, as well as those forms of mystical experience induced by drugs—and many of them are—tend to be aware of this beyond within. One of the popular books of the 1960s about drug-induced mysticism is even called *The Beyond Within*. But it is the parallel and complementary feeling of a "beyond without" that is especially interesting for the purposes of this book, since this feeling or mystical apprehension is now being duplicated rationally in quantum theory. It is fair to say that even those mystics who observe a beyond within say they also experience the melting away of the beyond without. This gives rise to their description of an undifferentiated unity.

However, as referenced, while mystical experiences have provided at least some motivation for the establishment of religious movements, there are good reasons for believing that the mystical experience in itself is a not a religious phenomenon at all, and its connection with religion is subsequent and perhaps even adventitious.[11] Discounting the various interpretations of the experience, which identify it with God, the Absolute, or the soul of the world, what you have left is basically an overwhelming emotional experience.

As I will discuss later, the same or similar experience can apparently be had with the aid of certain drugs. But, for those within a religious tradition, the experience is often described as relating to that tradition. In the theistic religions—Judaism, Christianity, Islam—the experience of the undifferentiated unity is expressed as union with God. But quite different interpretations of the same or similar experiences are given by those from different religions and cultures. And, of course, modern theoretical physics gives quite different explanations again, expressing the predilections of a secular, science-oriented, Western culture.

So, whether this "apprehension of an undifferentiated unity" is perceived via a classic mystical tradition of East or West, via drugs, or by modern quantum theory, what are we to make of it? Does it point to some reality or truth, or is it a crock, or *fantastica fornicatio* as St. Augustine would say? A fundamentalist skeptic, most certainly a cynic, and probably many current scientists would say that the mystical experience, or mystical consciousness, is completely subjective and relates to nothing outside itself. That it is, indeed, *fantastica fornicatio*. But, then again, one could say the same thing about reality as described by quantum theory. Since nobody, including scientists, knows much of anything about the nature of the human mind and human consciousness, to say that mysticism—or anything else—is subjective is not to say very much at all. Everything, including reality itself, may be subjective.

But there is a huge body of at least circumstantial evidence from mystics, theologians, philosophers through the ages, and scientists that sometimes and somehow some of these experiences, some of these theories, do indicate a reality or truth beyond what we experience and comprehend.

We perceive only that which our organs of perception are capable of perceiving and our minds are capable of translating and understanding. This ends up being only a small fraction of what is *actually there*. It is, therefore, easy to be deceived. In fact, it's practically axiomatic that we will be deceived. Our vision is selective. Each of us is selective of our experience in different ways, depending on heredity, environment, mental capability, custom, et cetera.

Through our consciousness we experience two worlds: one outside ourselves and one inside ourselves. We explore and try to understand both worlds. Using our reason, we are able to consider and interpret experiences, whether from outside via sense perception or from within. But the fact of experiences is primary. Without them we would have nothing to reason about. However, we have great leeway—I'm not sure I would exactly call it freedom since we are well conditioned by the time we can think rationally—in our ability to decide on the priorities and meanings we attach to our experiences. Most of us, especially in the West, put most of our faith in our external experiences and attach validity and truth mostly to them. This is the history of science and remains a focus of science. But others, notably mystics, put more validity and truth on their interpretation of their internal experiences. Now, these two worlds—the outer world of external experience and the inner world of internal experience—are beginning to be mingled, and therefore confused, by quantum physics.

Essentially, whether we pursue the uncovering of truths by looking outward to the physical world or inward to one's self-consciousness, the direction we choose is based on our particular faith perspective. In order to know at all, we must have faith. We may have the faith of the scientist, that of the mystic, that of the committed Christian or Buddhist, or indeed the faith of the faithless, but we start from a faith perspective, a conscious perspective that we choose from several possibilities.

St. Augustine wrote: "Understanding is the reward of faith. Therefore, do not seek to understand in order that you may believe, but make the act of faith in order that you may understand; for unless you make an act of faith you will not understand." A thousand years later Nicholas of Cusa wrote similarly: "In every science certain things must be accepted as first principles if the subject matter is to be understood; and these first postulates rest upon faith." Finally, in more recent times, astronomer Sir Edward Appleton said:

> So I want to make the assumption which the astronomer—and indeed any scientist—makes about the universe he investigates. It is this: that the same physical causes give rise to the same physical results anywhere in the universe, and at any time, past, present, and future. The fuller examination of this basic assumption, and much else besides, belongs to philosophy. The scientist, for his part, makes the assumption I have mentioned as an act of faith; and he feels confirmed in that faith by his increasing ability to build up a consistent and satisfying picture of the universe and its behavior.[12]

The nature of our faith—how we tend to think—is in large measure the result of the environment in which we live. In the West, our roots connect to classical Greece, where logical reasoning was developed to a very high level in natural philosophy, or science as we now know it. Science has been perhaps the single greatest achievement of the West. But because logical reasoning has been so successful in the rise and development of science over the decades since Galileo and Newton, we in the West tend to view it as the only valid method of gaining knowledge. As referenced earlier, the huge influence of Sir Isaac

Newton virtually enshrined this as the only way of gaining true knowledge. However, even in science, it was never the only way. As well as the path of discursive reason (i.e., moving from a premise to a conclusion in logical steps), there has always also been the path of intuition, creative insight, imagination—call it what you will.

Chapter Eight
Mysticism and Intuition

We learn from Plato that there are different levels of knowledge that are accessed in different ways, such as discursive reasoning and contemplation. For those within the Christian tradition, there is also revelation. Various paths have been recognized not only by poets, mystics, and philosophers but also by scientists. One of the founders of quantum physics, Max Planck, wrote in his autobiography: "When the pioneer in science sets forth the groping fingers of his thought, he must have a vivid, intuitive imagination, for new ideas are not generated by deduction, but by an artistically creative imagination."[1] A combination of intuition and discursive reasoning is often the basis of great breakthroughs in science. And great intuitive breakthroughs in science, as in music, or art, or spirituality, or anything, come to those who are thoroughly schooled in the subject matter.

A brilliant intuitive breakthrough in science comes to a scientist, not to a musician, just as a breakthrough in music comes to a consummate musician, not to a scientist. But the breakthrough itself seems to transcend one's knowledge base, certainly one's perceived base. One appears to connect to something, a truth perhaps, that is beyond one's immediate reality, if not oneself. It appears to connect them to a certainty—which is always beautiful—that they could never be connected with via discursive reasoning alone. But this experience tends not to have any ecstatic element, at least not in the sense of any "apprehension of an undifferentiated unity." However, it is similar, within von Hugel's comparatively broad definition, to a classic mystical experience; as being, not everything in any one soul, but something in every soul, and presenting, at its fullest, the amplest development of certain special natures. For the musician this then has to be communicated via notation, for the scientist, by a formulated theory that can be tested through discursive reasoning and/or by mathematical formulae, and ultimately, ideally, by controlled experimentation.

A good example in science is Heisenberg's discovery of the uncertainty principle. This is described by Shimon Malin and begins with an outline of Heisenberg's early experience after the end of World War I. The city of Munich, Germany, was in utter confusion. Heisenberg joined a volunteer force to help restore order. In his spare time, he would retire to the roof of the Training College to read Plato's dialogues. It was here that he discovered the *Timaeus*, Plato's account of the creation and structure of the universe. Its detailed account of the geometrical shapes of Plato's atoms didn't make much sense to him. He was disappointed that someone of Plato's reputation would have succumbed to such fantasies. However, because of the continuing political disturbances in Europe, the eternal order described by Plato kept attracting him, especially the key question of the relationship between the eternal order and the temporal order (i.e., the relationship in Plato's terms between the "intelligible world," the world of Forms, and the order in the visible universe in which we live). This preoccupation with the idea of order led him to a profound experience.[2]

A few months after he read the *Timaeus,* Heisenberg was part of a large gathering in a castle courtyard. As he listened to the different speakers talk about different types of order, it became obvious to him that their suggestions clashed with each other in a decidedly disorderly fashion. In Heisenberg's own words:

> This, I felt, was only possible because all these types of order were partial, mere fragments that had split off from a central order; they might not have lost their creative force, but they were no longer directed towards a unifying center.... The shadows of the courtyard grew longer, and finally the hot day gave way to slate-grey dusk and a moonlit night. The talk was still going on when, quite suddenly, a young violinist appeared on a balcony above the courtyard. There was a hush as, high above us, he struck up the first great D minor chord of Bach's Chaconne. All

> at once, with utter certainty, I had found my link with the center.... The clear phrases of the Chaconne touched me like a cool wind, breaking through the mist and revealing the towering structures beyond. There has always been a path to the central order in the language of music, in philosophy and in religion, today no less than in Plato's day and in Bach's. That I now knew from my own experience.[3]

It is difficult to know with any precision exactly how Heisenberg was influenced by Plato's dialogue. Perhaps the *Timaeus* helped free him from an obvious and simple attachment to the common realist view of the universe. Certainly, he was attracted by Plato's description of an eternal order. He writes: "I was enthralled by the idea that the smallest particles of matter must reduce to some mathematical form. After all, any attempt to unravel the dense skein of natural phenomena is dependent upon the discovery of mathematical forms."[4]

As we have seen, Heisenberg, with Niels Bohr and other founders of quantum theory, believed that elementary particles are not "little things" and that they can only be adequately described in mathematical symbols and equations rather than in words. In effect, he saw these smallest units of matter as similar to Plato's Ideas, which are best spoken of in the language of mathematics. Certainly his early contact with Plato motivated him to look for a broader interpretation of reality and some connections between everyday events in spacetime and various fields of potentialities. In any event, in this particular setting, his contemplation of this work by Plato, combined with the music, gave him an experience of certainty that was beyond any discursive reasoning. He had found a link to a central order, however that might be interpreted. He says that he knew something "with utter certainty" but that he could not say what he knew.

This was a creative, mystical experience that had nothing to do with his background as a scientist. He had been listening to different speakers talking about various types of order. He had a feeling that these were only partial, disconnected from a central order, but he was unable to discern a way through all the conflicting opinions to anything central. This process was a process of discursive reasoning. But then something happened as the violinist began to play. Suddenly, Heisenberg found—really, experienced—a link to this central order. This illustrates a fundamentally different method of knowing. And yet when Heisenberg did think about this experience, he interpreted it in the light of his background and learning as a scientist. His interpretation was done in the cold light of logical discursive reasoning many years after the event had occurred. Heisenberg was a critical and rigorous thinker, but he believed that he had experienced a link with a deep truth that lay beyond even his discursive reasoning powers. This experience motivated and guided his subsequent thinking, but the experience itself was beyond thought.

A similar creative insight beyond any ability of

discursive reasoning was the specific motivation for Heisenberg's discovery, or realization, of the uncertainty principle. During his stay at Niels Bohr's institute in Copenhagen, he was trying to resolve problems relating to the mathematical formulation of quantum mechanics and some experimental evidence that appeared to contradict it. He tried for months to resolve this, with no luck. Then one evening all became clear. What motivated this clarity was due to a previous conversation with Einstein, specifically his words: "It is the theory which describes what we can observe." He was immediately, intuitively convinced that the key to the gate that had been closed for so long was here.

After walking for a while and thinking about it, Heisenberg realized that the experiment in which the path of an electron in a cloud chamber could be observed was perhaps not what it appeared to be; it was something much less. He thought to himself that

> in fact, all we do see in the cloud chamber are individual water droplets which must certainly be much larger than the electron. The right question should therefore be: Can quantum mechanics represent the fact that an electron finds itself approximately with a given place and that it moves approximately with a given velocity, and can we make these approximations so close that they do not cause experimental difficulties?

When he returned to the institute he did some calculations, which showed that this could be represented mathematically and that the approximations are governed by what was later called the uncertainty principle of quantum mechanics.[5]

For my purposes, what is important and relevant about these intuitive breakthroughs of Heisenberg, while mystical only under the broader definition, is that they are indicative of a way of knowing that is very different from the usual logical reasoning generally associated with science. They are far closer to what we generally associate with the arts or religion or the mystics. But they are as common to great creative breakthroughs in science as to any other field, and the breakthroughs always relate directly to the expertise of the person. Following his breakthrough experience, Heisenberg thought more about it and engaged in calculations before coming to understand what it all meant and before he was finally able to formulate the uncertainty principle.

This is a good example of how the original intuitive, even mystical, experience gets translated into the cultural paradigm of the person who experiences it. This should not be surprising, for how else would they be interpreted and understood? These insights often have the appearance of something given to the person, a sort of revelation coming from beyond oneself, even though they always appear to come to the prepared mind. The mind, often when in a tranquil state, makes a sudden leap and what was once obscure becomes clear.

This appears to be the case for many such original

creative experiences, whether by scientists, philosophers, religious innovators, or classical mystics. But there is also a common motivation: the search for reality or truth or God, within the faith that such actually exists. This common motivation most certainly connects science with all areas of truth seeking, whether they be philosophy, religion, or pure mystical union. "In that thou hast sought me, thou hast already found me," says the voice of Absolute Truth in their ears. Evelyn Underhill describes this as the first doctrine of mysticism. It could as easily be described as a first doctrine of modern theoretical physics, or indeed of science and metaphysics generally. She continues: "We seek. That is a fact. We seek a city still out of sight. In the contrast with this goal, we live. But if this be so, then already we possess something of Being even in our finite seeking. For the readiness to seek is already something of an attainment, even if a poor one."[6]

Given that both the mystics and the modern theoretical physicists want to escape the prison of the sense-world, transcend its rhythm, and attain knowledge of, or conscious contact with, a reality beyond the senses, is there a fundamental difference in their motivations? Evelyn Underhill, among other writers of the mystical experience, believes that there is. She believes that the desire to know more is the essence of science and the desire to love more is the essence of mysticism and religion. I believe that this is far too simplified an understanding. Rather, it appears to me that both the desire to know and the desire to love is evident in both;

in fact, the desire to know on the part of the scientist and the philosopher is a desire indistinguishable from the desire to love.

For example, the admiration of the theoretical physicist for the beauty of the math or the simplicity of the equation is a statement of love. Of course, one can say that the desire for knowledge is, at least in part, a desire for love. Passion is a longing to know, in the deepest and fullest sense, the thing adored.

> Love's characteristic activity is a quest, an outgoing towards an object desired, which only when possessed will be fully known, and only when fully known can be perfectly adored. Intimate communion, no less than worship, is of its essence. Joyous fruition is its proper end. This is true of all of Love's quests, whether the Beloved be human or divine.[7]

Or whether the beloved be a mathematical theory so beautiful and simple and perfect that it can be a theory of everything, a theory of reality, and a theory of truth.

Yet the methods of mysticism and of science are fundamentally different. One is primarily emotional, the other primarily intellectual. The classical mystic, through contemplation and often various types of mental and physical exercises, looks inward and connects emotionally to what has been described as an undifferentiated unity. While there are intellectual elements, it is fundamentally emotional beyond the intellect. The act of contemplation may be described as a sort of psychic gateway, a method of moving from one level of consciousness to a higher level. It is the condition under which the mystics shift their field of perception and obtain their new and characteristically mystical outlook on the universe and reality.

Contemplation serves to release the mystics' so-called transcendental sense and allows them to move from one level of consciousness to another. The scientists, even the theoretical quantum physicists—while they are emotional and certainly contemplative and may look inward for mathematical intuitions—fundamentally look outward, incorporating the insights and tests of others and putting these and their own to the test of others; an essentially intellectual exercise with intuitive and emotional elements. Their motivations and even their creative breakthroughs may be similar. But their processes are fundamentally different.

The individual discipline of the mystics is to arrive at a stage where the self melds with the unity that is reality, where there is no longer a self. This is a spiritual world, a spiritual reality, that they believe should be the only goal of "pilgrim man." They believe that although we live in and usually believe in a world of appearances, we are not hermetically sealed from this spiritual reality. They believe that the "mystical path" is the way in which a dedicated individual may connect directly with it. Those who have the necessary faith and dedication to take this path may receive the truth at the end of it: experience Absolute Life, Absolute Beauty,

Absolute Truth, beyond time and place. This news often gets translated, sometimes poorly, into the language of religion.

In theological language, the mystics tend to believe that the spirit of man is essentially divine and is therefore capable of immediate communion with the divine. Very different from science or philosophy or theology, the mystics theorize, and perhaps even prove in their own way, the existence of an absolute and the possibility of knowing and finally attaining it. They deny that knowledge is limited to sense impressions, to the intellect, or in any way to the content of normal consciousness.

In mysticism, that love of and pursuit of truth or reality, which is the beginning of all philosophy and science, leaves the sphere of the intellect and takes on the more, at least personally, assured sphere of the passionate and the emotional. Where philosophy and science guess and argue, notwithstanding creative breakthroughs of genius, the mystic lives and experiences and communicates in the language of firsthand experience.

"Oh, taste and see!" they cry, in accents of astounding certainty and joy. "Ours is an experimental science. We can but communicate our system, never its results. We come to you not as thinkers, but as doers."[8]

Chapter Nine

Mysticism as Primarily an Eastern Phenomenon

Mystical experiences and the philosophic speculations that surround them—while there are many similarities—have usually been interpreted differently in the religions of the East and the West. A common feeling of mystical union with the universe, which is often described theologically as either monism (all spirit is one, or the soul of man is not distinct from God) or pantheism (all is one, or God is everything and everything is God) and which is common to religions of the East, was at odds with the theology of Judaism and Christianity and Islam, which believes that God is always separate and distinct from his creation, that one is never a part of him. There is no large One.

Great Christian mystics (such as Meister Eckhart, Henry Suso, and John Tauler, who lived and worked during the fourteenth century in what is now Germany) would certainly have been familiar with the ecstatic state, and they were accused of incorporating a monistic philosophy into their understanding of reality. However, the mystical union they desired, they said, was a new life in a union of love with God, so they had to interpret their experience and their hope in light of their tradition. As noted earlier, some writers on the mystical tradition say that fundamental differences in the mystical experience, even the ecstatic experience, reflect the religious and cultural setting and the expectations of the particular mystic.

As an aside, it is interesting to note that because theoretical physics now understands that matter and energy are interchangeable and that, at its most basic level, reality may be energy, the concept of monism should probably now be incorporated into pantheism.

The fourteenth century in Europe has been called the mystical century par excellence;[1] that is, of course, Christian mysticism. This century is considered unique in the history of mystical religion because of the extraordinary extent of the flowering of the human spirit.[2] While there were other Christian mystics before and after, I have referenced just the three big ones: Eckhart, Suso, and Tauler. These monks were the first to break away from a long-cherished mode of thought and substitute a new, and perhaps more profound, view of the relations between God and the universe. It was a view that let them—and with them Christian mysticism of the West—be a part of, or at least connected with, the common mystical experience and tradition. This view consisted of substituting the idea of the immanence of God in the world for the idea of the emanations of the world from God.[3]

The emanation theory supposes a radiation from above, while the theory of immanence assumes a self-development or manifestation of God from within. The pyramid could be considered the symbol of the first; the sphere being a symbol of the second. The theory of immanence declares God to be present everywhere and that man will realize heaven or hell in the present moment, denies that God is nearer the other side of the grave than this, equalizes all external states, breaks down all steps and partitions, and has man at once escape from all that is not God and so find only God everywhere.[4]

Evelyn Underhill also deals with these two very different ways of looking at the relationship between God and man. She explains that the theory of emanations declares God's utter transcendence. God, or the Godhead, is conceived of as removed by a vast distance from the world of sense. While our world was generated by the Godhead, the Godhead can never be discerned by man. When this theory of the Absolute is accepted, the movement of the soul toward union must be a journey upward and outward through a long series of intermediate states or worlds. She goes on to explain that to the holders of the theory of immanence, the quest for the Absolute is no longer a journey but a realization of something implicit in the self and the universe. She uses the phrase "the Spirit of God is within you."

Meister Eckhart used language appropriate to both the theories of emanations and of immanence.[5] In point of fact, Eckhart has been described both as a pantheist[6] and as a monist.[7] Either of these is a reasonable assessment, given that he was accused of heresy by the Church for making an identification between creature and Creator. However, other writers take a more conservative position. One concluded that because Eckhart believed that God created the universe out of nothing—that he called it into existence from the void, so to speak—his belief was opposed to pantheism.[8] Another believed that, while Eckhart is intellectually drawn toward a semipantheistic idealism, to be a true pantheist one must believe that all is equally divine, good or bad, and Eckhart could never have countenanced such a theory.[9]

It is clear that Eckhart and the other Christian mystics in the West experienced something within the mystical experience that was the same as or similar to mystics everywhere, and they had to deal with it within their everyday lives, in the "real world." They had to interpret it, as best they could, within the intellectual understandings that they possessed and within the traditions in which they lived. At the very least, the theory of immanence provided a theological rationale for these Western mystics to accept the emotional reality of the mystical ecstatic experience within the tradition of Christianity, and while never accepted generally by the Church, mysticism served its purpose. Mysticism can be considered a legitimate, though minor, part of the Western religious tradition.

That said, however, while the ecstatic experience does appear to be common wherever and under whatever conditions it occurs, the religions of the East, historically and currently, give unqualified expert testimony to this experience. Mysticism and the mystical tradition have been and are central to Eastern religions, and the theology that speaks of "One" and "undifferentiated unity" and that accepts monism and pantheism is common there. That is probably why it is easier to connect the insights of quantum physics to the insights of the East than to those of the West.

To help in this regard, what follows immediately is some background on the Indian side of these Eastern traditions. Looking, however briefly, at the origins of Hinduism and Buddhism is sufficient to provide some appreciation of the central characteristic of the Eastern outlook on reality. It also provides an appreciation of the fact that in the Eastern tradition as in the Western from the Greeks, thinkers, philosophers, mystics, and religionists of all stripes have always questioned and doubted the reality of the world we perceive around us.

Sometime around 1500 BC the country now known as India was invaded from the northwest, by tall, light-skinned people of Indo-European stock, called Aryans, who came over the passes of the Hindu Kush mountains. The struggle with the indigenous dark-skinned Darvidians was continuous. These wars and intertribal clashes over the years were later immortalized in the great Hindu epics: the Ramayana and the Mahabharata.

As the Aryans settled down, their naturally adventurous and wandering spirits turned inward to the imagination. Folk tales and epic stories were developed, and the hymns and prayers of their priests gave voice to their religious conceptions. Out of these last, together with ancient spells, came Hinduism's earliest sacred writings: the Rig Veda, Sama Veda, Yajur Veda, and Atharva Veda. Virtually all knowledge of the religion of the Vedic Age comes from a study of these works.

The Rig Veda (meaning "stanzas of praise") is an anthology of religious poetry in ten books. It was not committed to writing until about the eighth century BC. The poetry was deeply affected by the apparently mysterious, awe-inspiring forces of nature. All nature was looked upon as a living presence, an aggregate of animated entities—in this case, a large number of Rig Vedic gods. In time, as the sacrifices to these various gods developed into elaborate ceremonies, priests, each with specific duties, took charge. Again, over a long period of time, the most important priest came to be known as the Brahmin, or presiding priest, who offered the central petition or *brahma* (prayer).

By around the end of the seventh century BC, the Aryan occupation had resulted in a somewhat familiar stratification of society. The Aryans, the conquerors, occupied the upper strata of the social order, with the dark-skinned Dravidians below them. While the separation between classes was not yet solidified at this early stage, four distinct social groups—the Kshatriyas or nobles, the Brahmins or priests, the Vaisyas or Aryan common people, and the Shudras or non-Aryan blacks—were developing. This was, at least partially, based on color or varna, the Hindu word for caste.[10]

By this time, the Brahmins had evolved into a

very powerful caste. They were the key ingredient, the connecting link, in a process that reached all portions of the universe: hell, heaven, and earth. It was believed that they altered the very course of the cosmos through their sacrifices. This power came primarily from the utterances of sacred prayers in connection with the sacrifices. They made the claim that if they performed the sacrifices properly, as prescribed in certain treatises, which were called the Brahmanas and which they appended to the Vedas, they could procure the desired results. However, mixed with directions for sacrifices, there developed a growing sense of unity with the universe. Some of the priests speculated that if the holy power that worked through the prayers could alter the course of cosmic order, then that power, because it controlled both gods and men, must be ultimate. Could it perhaps be the true central power in the universe? Could this ultimate reality of the universe be called Brahma?

While this deep philosophizing is generally thought of as occurring with the stratification of society and in these circumstances, it was also a small part of the earlier literature. It has been generally thought that the Rig Vedic religion is essentially polytheistic, taking on some pantheistic aspects in only a few of its later hymns. However, a deeply abstract philosophizing turns up in some hymns as a reminder of its transition through naturalistic polytheism (belief in and worship of many gods) to monotheism (belief that there is only one God) to monism (belief that all spirit is one and the soul of man is not distinct from God). The evolution through monotheism to monism rather than ending with monotheism is interesting for us in the monotheistic West.

This transition was effected because of the growing influence of the concept of what was known as *rita*. Rita was originally the path of the Zodiac, within which the apparent motions of the divas (luminaries or gods) are confined. The divas are said to be born in rita and governed by it. Rita then came to denote the cosmic order or law that governed nature. In the moral world, the word *rita* designates "order" through the meanings of "truth" and "right," and in the religious world, "the order" takes the form of sacrifice and rite. So the way to the later conception of the Absolute, which is impersonal and designated by the neuter term "Brahman," had been paved by the abstract concept of rita.[11]

With the development of the Brahmanas, one of the great speculative eras in the history of religion was opened. It was a speculation that was echoed—or perhaps paralleled—in the Western philosophical tradition and is certainly basic to the mystical traditions of East and West, as well as to finding a modern outlet in much of the speculations of theoretical physics. This speculation pressed on to new and philosophically profound interpretations of reality. The writings, which were done over several centuries and terminated around 700 BC, presented these new ideas as appendages to the Brahmanas. They form the famous treaties known as the Upanishads. For the study of the religions of India, and certainly for an understanding of the growth of the mystical and philosophical thoughts that we have been examining, this is the most important series of speculations.

The word *Upanishad* is derived from the root *upa ni sad*, which means "to sit down near someone." This probably refers to a pupil sitting down near his teacher for instruction. Over time, the word incorporated a sense of secret communication or doctrine that was imparted at these sittings. Still later, the word came to be applied to the texts that incorporated such doctrines. The Upanishads are usually called Vedanta, or "the end of the Veda," because they came at the end of the Vedic period, they were taught at the end of Vedic instruction, and the late philosophers found the final aim of the Veda in them. It is thought that the Upanishads are in fact the legitimate development of the skepticism found much earlier in the Rig Vedic hymns. The most important contents of the Upanishads are found in the philosophic speculations.[12]

The philosophers of the Upanishads, like current theoretical physicists and philosophers in the West, explore the ultimate truth behind the world of creation. There is no consensus in them, but they tend toward monism in dealing with ultimate questions such as: What is reality? Of what is the universe the expression? Is it real in itself? Or is it merely an appearance, even an illusion? How did human experience, illusory or real, come to be? What is the meaning of human life?

While there may be no clear consensus, the Upanishads generally agree on a fundamental theory. The basis of all being, whether material or spiritual and in whatever form of the cosmos and beyond it may take, is an all-inclusive, unitary reality that is beyond sense-apprehension, ultimate in substance, infinite in essence, and self-sufficient; this unitary reality is the only really existent entity. This reality is most commonly called Brahma or Brahman. There are no precise definitions, and descriptions vary. However, it is clear that the Brahma is all that is objective, the entire world that is given to our senses, and all that exists outside of us; everything is ultimately phases of the One. But Brahma is also everything that is subjective, the entire inward world of feeling, and self-consciousness. The term for this inner self is *atman*. Many of the Upanishads say that there is a clear identity between Brahma and atman.[13]

Thus, the Upanishads variously express their findings in the identity of Brahman—that highest principle that manifests itself in the various forms of creation and that receives everything back at the time of dissolution—and atman, which is the individual self. The goal is to achieve the union of the individual soul with Brahman. This cannot be achieved through work. The only way is to realize the identity of the soul with Brahman.

Brahma is atman. The true self of a man and the world-soul are one; they are identical. The All-soul is the stuff—sometimes considered mental and sometimes material—of which the human soul and its consciousness are formed, and there is no distinction between the two. Therefore, Brahma, (the objective All) and atman (the subjective or particular self) can be equated, and the ultimate reality is Brahman-atman.

> If one single doctrine were to be selected from the old Upanishads as representing the quintessence of Upanishadic philosophy, and if we were asked to sum it up in one sentence, that sentence would be: "The universe is Brahman, but Brahman is the atman." This unity of Brahman (the cosmic principle) and atman (the psychic principle) is expressed in the Upanishadic dictum "Thou art That."[14]

The philosophers of the Upanishads recognized that the subjective and the objective are one. In doing so, they took a major step that carried their philosophy into mysticism. They thought of this recognition as entering nirvana. This step was taken by recognizing that when the human soul knows its complete identity with Brahma it celebrates this knowledge with a feeling of unity-approaching ecstasy. The experience of such assured knowledge was acknowledged to be so beatific as to be indescribable and blissful. The technique, which of course came after many years of rigorous study and discipline, involved sitting and meditating in profound quiet of mind, seeking to know, verily know, not to have an opinion or a mere belief, but to be spiritually absolutely certain. Deep dreamless sleep comes nearest to providing an analogy for the state of union with Brahma.[15]

This perspective on reality carried over into the new religion of Buddhism, which arose out of Hinduism. As referenced, long before the period of the Brahmanas, the creative thinking that gave rise to the rich philosophical speculations of the Upanishads had already evolved in certain Rig Vedic hymns that express doubts regarding the efficacy of priestly cults and the belief in gods. It was probably these nonpriestly groups opposed to the Brahmanic works that formed the chief recruiting ground of the forest hermits and wandering ascetics who disconnected from the sacrificial ceremonies of the Brahmanas, renouncing the world and following the "way of knowledge." Buddhism represents, very probably, one fruit of this protestant activity.

Buddhism was founded in northern India in the sixth century BC by Siddhartha Guatama, who became known as "the Buddha" or "the Enlightened One." Following many years of physical and philosophical struggle, he achieved enlightenment as part of an ecstatic mystical experience while meditating for a lengthy period under a bo tree. Notwithstanding this apparently mystical experience of entering nirvana, the original teachings of the Buddha reject both speculative philosophy and religious devotion. Rather, his mystical insight emphasized a more practical need to control one's desires and live right. However, as the movement developed into what is described by John Noss, in *Man's Religions*, as a "family of religions rather than just a single religion," its many parts incorporated the speculative philosophy and mystical traditions of its Indian roots.[16]

This was especially true of its largest and most elaborate branch, the Mahayana, which arose sometime between the third century BC and the first century AD and spread aggressively throughout the East. Its more conservative branch, the Hinayana, never had the same popular religious appeal. The Mahayanists, reflecting the influence of Brahmanism, perceived a reality behind all phenomena that displayed itself over and over in recurring events.

This ancient understanding coming out of India has many historical and current equivalents. The feeling, the certainty, the bliss, the ecstasy that comes when a subject reaches a certain mystical state is common. So too is the feeling of oneness with the universe or with a reality that is behind all appearances. It is a feeling of unity and that the subjective and the objective are one. As we have seen, this is also a common perspective of theoretical physics; not the mystical state, but a belief in the possibility that all may be one and in the sense that the basis of all may be energy, and most certainly that the world of our senses is not reality.

It is these connections of the mystical traditions, principally of the East but not excluding those of the West, with modern quantum theory that are the subject of the book *The Tao of Physics* by Fritjof Capra. A book by the Dalai Lama, *The Universe in a Single Atom*, makes similar connections between Buddhism and quantum theory.

Chapter Ten
Connecting the Mystical and the Quantum Traditions

In natural philosophy, from the time of the Greeks, the mass of objects has been associated with "stuff," indestructible material units that were called atoms by the Greeks but became smaller and smaller elementary particles of one type or another by modern theoretical physics. The question was whether one could divide matter again and again and finally arrive at some smallest indivisible unit, or whether, at its most fundamental level, matter turned out to be somehow immaterial. Rather than "stuff" or "little things," relativity theory shows that mass can be energy. Energy can morph into matter, and matter into energy. It is this understanding, as well as the many other mysterious elements of quantum theory, that has led to associations with mystical insights.

Physicist Fritjof Capra says that the worldview of modern physics has shown that the idea of "basic building blocks of matter" is no longer tenable. In the past, this concept was very successful in explaining the physical world as made up of atoms, then the structure of an atom as a nucleus surrounded by electrons, and then the structure of the nucleus in terms of two nuclear building blocks, the proton and the neutron—"elementary particles." However, when they were examined closely enough, the proton and neutron turned out to be composite rather then elementary structures, forcing physicists to hope that the next generation of constituents would be the final and ultimate components of matter. And on and on, ad infinitum.

In the meantime, relativity and quantum theories made the existence of units or building blocks of matter increasingly unlikely. They showed that energy of motion can be transformed into mass and suggested that particles are processes rather than "little things."[1] The discovery that mass is simply a form of energy has modified our concept of a particle; now not considered to consist of "stuff" but as a bundle of energy. Since energy is associated with activity and process, the nature of subatomic particles is dynamic. These dynamic "energy bundles" form the stable nuclear, atomic, and molecular structures that build up matter and give it its macroscopic solid aspect. Quantum theory has shown that particles are not isolated bits of matter, but are probability patterns that change into one another, interconnected in an inseparable cosmic web.[2]

Capra believes that the creation of material particles from pure energy is the most spectacular effect of relativity theory. After relativity and high-energy experiments involving the collision of subatomic particles, the whole question of what is real appeared in a new light. The experiments showed the dynamic and ever-changing nature of the elementary particle world. All particles can be transmuted into other particles. They can be created from energy and can vanish into energy. In this subatomic world, the whole universe is a dynamic web of energy patterns. Thus modern physics sees the universe as a dynamic, inseparable whole that always includes the observer in an essential way.

> The exploration of the subatomic world in the twentieth century has revealed the intrinsically dynamic nature of matter. It has shown that the constituents of atoms, the subatomic particles, are dynamic patterns which do not exist as isolated entities, but as integral parts of an inseparable network of interactions. These interactions involve a ceaseless flow of energy manifesting itself as the exchange of particles.... The particle interactions give rise to the stable structures which build up the material world, which again do not remain static, but oscillate in rhythmic movements. The whole universe is thus engaged in endless motion and activity; in a continual "cosmic dance" of energy.[3]

This is the world of mysticism, of Eastern religions, and is a major reason why many physicists are worried about and frightened by this understanding of quantum mechanics. And why many of them, I am sure, are not at all pleased by Capra's book. Perhaps because he is not a physicist but rather a philosopher used to very nuanced discussion and debate about truths, the Dalai Lama writes about these things much more tentatively than does Capra. But what he says supports the argument that what quantum theory says about the world and reality is very similar to what the mystics and Eastern religions—in his case Buddhism, of course—have been saying for ages.

The Dalai Lama explains that one of the deep philosophical insights of Buddhism comes from what is called the theory of emptiness. This theory recognizes that a deep disparity exists between our perception of the world—including our existence in it—and the way things really are. We relate to the world and to ourselves as if these entities possessed some definable, discrete, and enduring reality, and we tend to believe in an essential core to our being—a discrete ego independent of the physical and mental elements that make up our existence.

The philosophy of emptiness says that this is a fundamental error. Any belief in an objective reality based on the assumption of an independent existence is illusion. It says that all things and events, both mental and material, and even concepts, such as time, have no objective, independent existence. Rather, everything is made up of dependently related events, of continuously interacting phenomena with no fixed, immutable essence. Everything is in constant and changing dynamic relations. So things and events are "empty" in that they do not possess any immutable essence, intrinsic reality, or absolute "being" that provides any kind of independence.[4] Our commonsense, or naïve, view of the world leads us to believe that things and events have some enduring reality and that the world is real. According to the theory of emptiness, it is not. It is illusion.

As referenced in the introduction to this chapter, a principal motive of philosophical and scientific inquiry since the time of the ancient Greeks has been to find the irreducible building block of matter: the "littlest thing." Once it was demonstrated that the atom could be divided, the search was on for the smallest elementary particle that could not be divided. That search is ongoing. As the Dalai Lama points out, Buddhist thought argues that this scientific search for the irreducible "little thing" is misguided.[5]

In this sense, as in most senses, Buddhist and mystical thought generally falls on the side of the Copenhagen school of quantum theory and schools of thought aligned with this approach to the concept of reality. Both are convinced that one cannot consider subatomic particles to be mutually exclusive entities. These elementary constituents of matter can be considered either particles or waves, and it is the experiment and the experimenter that determines which. Heisenberg's uncertainty principle says that knowing an electron's exact position means that we cannot know its exact momentum, and knowing its exact momentum means that we cannot know its exact position. This shows that the observer is a participant in the reality being observed.

The Dalai Lama says that this has long been an important principal in Buddhist thought. The recognition of the fundamentally dependent nature of reality—that matter and mind are codependent, called *dependent origination*—lies at the heart of the Buddhist understanding of the world and the nature of human existence. The world is made up of a network of interrelations. The notion of an observer-independent reality is untenable.[6]

So how does one resolve this dilemma between our commonsense view of the reality of the world around us with the view of Buddhism, of other Eastern religions, of essentially all mystical movements, and now of modern quantum physics that this commonsense view is an illusion? The religions and mystical movements have resolved this, at least to their own thinking, in a variety of ways. The Dalai Lama explains that Buddhism has done this via the notion of two truths, the "conventional" and the "ultimate," that relate respectively to the everyday world of our experience and to things and events in their ultimate mode of being, which is on the level of emptiness.

Now, it seems to me that this theory is simply a practical solution that summarizes the dilemma, rather than resolves it. To rephrase, there is the reality or truth that we experience via our senses, which isn't really a truth at all, and then there is the real reality or truth that is beyond our experience, or at least beyond our practical, rational sensory experience. In point of fact, it looks like there is no resolution at all.

Similarly, quantum physicists try to deal with the dilemma in their own practical ways, but not yet in any one practical way. There is the theory of emergent properties proposed by Robert Laughlin. It says that the reality of the quantum world emerges in some magical, or at least so far unknown, way into the reality of the conventional world that we experience. Maybe the discovery of a "God particle" will help resolve this. The Dalai Lama says that most quantum physicists appear to relate to their field in a schizophrenic manner. When they work in their laboratories and struggle to find and identify "little things," they are realists. They talk about elementary particles moving around from here to there. However, when you change the topic to a more philosophical discussion, most would say that nothing really exists without the apparatus that defines it.[7]

What is obvious is the truly paradoxical nature of reality or truth revealed by modern physics but also for centuries by Buddhist philosophy. While the Dalai Lama doesn't deal with broader issues, we need to emphasize again that this paradoxical nature of reality and truth has been an important part of all religions and philosophies of both East and West.

Besides its more obvious technological benefits, the most important philosophical benefit—if we can call it a benefit—of quantum physics is that science, our universally accepted path to secular truth in the West, is now questioning not only what is real but whether it is even possible to know what is real. These are deep metaphysical questions, previously thought appropriate only for mystics and philosophers. These questions may be becoming, tentatively, mainstream.

Fritjof Capra firmly believes that modern physics has confirmed one of the basic ideas of all mysticism: that the concepts we use to describe the world around us are limited, that they are not features of reality, as we in the West have tended to believe, but creations of the mind. When we expand the realm of our experience, the limitations of our rational mind become apparent and we have to modify or abandon some of our concepts. He uses our notions of space and time to illustrate this point. These notions figure prominently on our map of reality, but relativity theory has shown them to be illusory. Eastern philosophy has always maintained that space and time are intellectual concepts that are relative, limited, and illusory.

> The central aim of Eastern mysticism (in fact, all mysticism) is to experience all phenomena in the world as manifestations of the same ultimate reality. This reality is seen as the essence of the universe, underlying and unifying the multitude of things and events we observe.... The mystics see the universe as an inseparable web, whose interconnections are dynamic and not static. The cosmic web is alive; it moves, grows and changes continually. Modern physics, too, has

> come to conceive of the universe as such a web of relations and (like the mystics) has recognized that this web is intrinsically dynamic.[8]

The Dalai Lama draws many similar connections between the theories of quantum mechanics and those of Buddhism. But he also raises some important general questions regarding the truths of science. These are good philosophical questions, applicable to all thoughtful people.

One centers around the origins of the cosmos. Both science and Buddhism are essentially nontheistic in their philosophical orientations, and neither postulates a transcendent being—God—as creating the cosmos or universe. However, if the Big Bang is taken to be the beginning, and unless one simply refuses, like many physicists, to speculate beyond this cosmic explosion, one is left with some kind of creative effect. This is not necessarily the god of the Judeo-Christian traditions, but it is some kind of transcendent principle—a godhead, if you will. However, as a counter to this necessary conclusion, some scientists now suggest that the Big Bang may have been less of an absolute starting point than simply one of perhaps an infinite number of thermodynamic-like instabilities. They are suggesting that the Big Bang may not have been the absolute beginning of anything.

As we have seen in chapter 5, understanding the exact beginning of the universe, the actual bang, is probably an impossible task. This instant was a singularity in which all mathematical equations and the laws of physics break down. Perhaps a so-called unified theory that brings together the insights of quantum mechanics and relativity will provide an answer to understanding the issue. Or perhaps not. Quantum mechanics has demonstrated that it is impossible to predict how even one elementary particle will behave in a given situation. One may only know the probability of how the particle will behave. If we can never arrive at a complete understanding and description of even a single atom, how can we expect to do so about the entire universe? Still, we can all speculate about it.

As the Dalai Lama says,

> even with all these profound scientific theories of the origin of the universe, I am left with questions, serious ones: What existed before the big bang? Where did the big bang come from? What caused it? Why has our planet evolved to support life? What is the relationship between the cosmos and the beings that have evolved within it? Scientists may dismiss these questions as nonsensical, or they may acknowledge their importance but deny that they belong to the domain of scientific inquiry. However, both these approaches will have the consequence of acknowledging definite limits to our scientific knowledge of the origin of the cosmos.[9]

Another question of the Dalai Lama centers on human

consciousness. He believes that because the experience of consciousness is entirely subjective and science is so thoroughly third-person in its methods, science has made very little headway in understanding consciousness, nor has science developed an adequate methodology for studying it.

Despite the reality of our subjectivity and centuries of philosophical examination, there is very little consensus amongst the scientific community as to what consciousness actually is. There are, of course, plenty of theories. At one extreme, behaviorism defines it in terms of external behavior, thereby reducing it to verbal and bodily action. At the other end of the spectrum is what is usually described as Cartesian dualism: the theory that the world is comprised of two independent and real things, namely matter and mind.

In general, Western science has attempted to understand consciousness in terms of brain function. This means grounding the nature and existence of mind in matter, or the physical makeup of the brain. Some theories state that mind and consciousness are no more than simple operations of the brain and that sensations and emotions are no more than chemical reactions. Thus, Consciousness is considered to be a special kind of physical process that arises via the specific structure and dynamics of the brain. However, this view that all mental processes are necessarily physical processes is—like many other scientific theories—a metaphysical assumption, not a scientific fact. If mind is simply reduced to matter, as it is in third-person experiments, then one cannot adequately explain the emergence of consciousness. "A model of increasing complexity based on evolution through natural selection is simply a descriptive hypothesis, a kind of euphemism for 'mystery,' and not a satisfactory explanation."[10]

While many scientists maintain a faith in the physical basis of consciousness, some question it. "Many differing views have been expressed with regard to the relation of the state of the brain to the phenomenon of consciousness. There is remarkable little consensus of opinion for a phenomenon of such obvious importance."[11] This subject is wide open for speculation, but also for serious study. "How does brute inanimate matter become conscious, or rather self-conscious? No one has any idea. Consciousness and matter are as different as chalk and cheese. Nothing in the material world gives a clue as to how parts of it (our brains) become conscious."[12]

This area of human consciousness and the limitations of science in understanding it is an important part of the Dalai Lama's book, *The Universe in a Single Atom*. He believes strongly that the third-person method, which serves science so well in so many areas, is inadequate to explain consciousness and that it must be integrated with a first-person perspective. This belief coincides with that of many theoretical physicists, who have been saying for some time that the third-person method is equally inadequate in dealing with the entire quantum world, where the subject and the object cannot be separated.

Chapter Eleven
Drugs and Popular Western Mysticism

For very good reasons, Western science mistrusts first-person accounts. This is probably one reason leading scientists so vociferously opposed quantum theory originally and why it is still so troubling to some. It shows clearly that you cannot separate the experiment from the experimenter, the examined from the examiner, the world from the human observer, or reality from human consciousness. So, one way or another, first-person perspectives have to be taken into account, even though such perspectives are questioned.

Since the advent of quantum theory, some theoretical physicists have concluded that all the theories of natural phenomena and all the laws of nature are creations of the human mind—first-person rather than third-person realities. However, while this has connected them to the worldview of mysticism, certainly the majority of physicists would be distinctly out of sync with such an association, even while struggling with the centrality of the first-person perspective.

The mystic way leads to certainty, to assurance. This is important psychologically and perhaps spiritually, at least for many people. But it can also be a disaster politically and economically, not to mention spiritually for many. While the emotional experience of mysticism—that of an undifferentiated unity—appears to be common, there is no agreement as to exactly what this is or what it means. There is no objective standard by which it can be measured or evaluated. Understanding of undifferentiated unity is dependent on the explanations of those who have experienced it and who say that it can never be adequately communicated by words. It is subjective in the most fundamental meaning of that term. It must be experienced, and once it has been experienced it can never be refuted.

Let me emphasize again that while scientific truths have obviously been and still are accepted as truth by many and indeed give a certainty similar to both religion and mysticism, these truths are always theories—approximations, if you will—of reality. They are always open to debate, new insights and discoveries, and continual skepticism. Nothing I have read about Eastern religions, religion generally, and certainly nothing about mysticism leads me to believe that there is anything anywhere near this kind of freedom for doubt, speculation, discovery, and growth.

Let me also emphasize that skeptics are simply doubters. They are not cynics. True skepticism is healthy, since it is obvious that we are dancing in the dark. But here, as everywhere, one has to be careful.

It is easy to go from being a healthy skeptic—one who continually questions truths—to one who denies that there are truths. How does one know this if one denies the possibility of knowledge? To go this far is to pass from true skepticism to certainty and then to dogmatism. One has simply traded one version of truth for another.

Skeptics continually question, but they can do so within a strong faith in and pursuit of reality and truth, however that pursuit is understood. With its roots in the Greek tradition, Western natural philosophy and science have raised skepticism to the center of their methods. Because of this they have been, and remain, movements of enlightenment and liberation. They aim to free the mind from the bonds of ignorance. They present the world as being accessible to human reason. This way, rather than through religious or mystical mysteries, freedom and the good life lie. It certainly has for the vast majority of us in the West.

Both modern science and mysticism take the view that everything in the universe is connected to everything else. The properties of any part are influenced, if not determined, by the properties of the other parts. Both physicists and mystics realize that it is impossible to fully understand and explain any phenomenon. But then they go their separate ways. Physicists are satisfied with an approximate understanding and a continual questioning and pursuit. Mystics are not interested in and certainly not satisfied by approximate or relative knowledge; they want absolute knowledge, absolute assurance. Theirs is an emotional rather than an intellectual pursuit of truth, and while some of their conclusions about reality may sound similar when they are explained, the two methods are completely different.

As there are no, or very few, safety checks or independent evaluations of the varieties of mystical experiences, the mystic way takes many paths. At the conclusion of chapter 8, I included a quote to describe the mystics not as thinkers but as doers. Well, doers have a way of looking at options, and the mystic path is no exception. Some spend years of study and meditation and often undergo physical deprivation. Others look for shorter and easier ways. The ecstatic mystical experience—or "altered state of awareness" reported by many as the "apprehension of an undifferentiated unity"—may come to someone naturally, in the sense of not being specifically or at least consciously planned for, but it may also be induced through a variety of actions, from yoga to breathing exercises to fasting to self-mortification to the use of certain drugs—sometimes in combination. Breathing exercises change the chemical composition of the blood and provide a focus for rhythmic fixation of attention. Fasting, self-flagellation, and other forms of mortification have been practiced not only to assuage guilt or prove devotion but also to enhance mental awareness; these can cause body damage that in turn can cause chemical changes.

Medical doctors and psychiatrists who have studied this phenomenon report that, from their specific medical (i.e., physical) perspective, these experiences are all a result of sensory deprivation. They all affect a particular part of the brain, which causes these mystical experiences. Sensory deprivation experiments were quite popular and studied extensively in the 1960s using isolation as a method. It was claimed that sensory deprivation poisons a nerve center in the brain or nervous system that causes hallucinations and other wild experiences, including the apprehension referenced above. The experiments were conducted, at least in part, because of the extravagant claims made by some of the more popular users of psychedelic drugs that they experienced the same or similar insights to those of the great mystics. Mystical apprehension solely by the use of certain drugs was a growth industry in the 1960s. Of course, the use of drugs has always been a part of the mystic way, though not for everyone and perhaps not for the truly great mystics.

As we have been exploring throughout this book, men and women are continually expressing, in a great variety of ways, a desire to rise above their everyday selves and understandings and achieve some higher insight or truth. So-called psychedelic drugs became the fashionable and popular way to do this in the 1960s. Essentially, and certainly compared to the classic mystic way, this was "no fuss, no muss." The leading "guru" of this mystical, quasi-religious movement was a Harvard psychology professor, Timothy Leary. He was acting in the tradition of other psychologists and students of human perception—such as William James and Aldous Huxley, who also tried out various experimental drugs on themselves in an effort to induce states that would lead to extraordinary lucidity and bring light to the mind's unconscious and creative processes.

Although I would argue, and have argued, that while their experiences may have been similar to the mystical and religious experiences of the great mystics—the "apprehension of an undifferentiated unity" comes to mind—they were fundamentally different. The great mystics of both the East and the West took years to achieve their experiences and then went on to great works as a result of these experiences. These 1960s mystics popped a couple of pills, had some sort of an experience, and then popped more pills.

Probably, the discipline, the faith, the background, and the culture were far more important than the experience itself, certainly in the long term. But for the purposes of this book, it is still interesting and worth further consideration that the actual reported experiences under these drugs were similar to the ecstatic experiences reported by the mystics—perhaps because it was indeed sensory deprivation—and that what they say they experienced is a "state of realty" that is at least similar to the reality described by quantum theory. For all of them, there is a reality or truth beyond what we experience. The differences are in the means of attaining or achieving or understanding this truth rather than the end itself.

Enemies of the use of these drugs, in the 1960s and today, called them "mind-distorting." Their proponents called them "consciousness-changing" or "consciousness-expanding" agents, arguing that they widened one's window to the world, as well as one's window to oneself.

As referenced, many techniques—including hard work and a passion for truth but also a wide variety of drugs—are available to accomplish some sort of consciousness alteration. In the West, we are most familiar with alcohol. In the Orient, opium, which is a narcotic, may be favored. If an enhanced alertness of heightened contact with the environment is preferred, stimulants such as tea, coffee, Benzedrine, or cocaine may be taken. However, states of delirium, sedation, or stimulation are not states that would normally induce the apprehension of anything approaching an undifferentiated unity. This and other dimensions of awareness are apparently possible, ranging from the most profound feelings of mystical union with the universe to terrifying convictions of madness and from ecstasy to despair.

Drugs that mediate these various apprehensions have various names. They are called hallucinogens by some or, together with the effects they produce, psychedelics by others. The word *psychedelic* is used most often and means "mind-manifesting." Some would say, on good authority, that what they manifest is illusion, which is described as an error based upon some sensory cue (e.g., a crack on the wall identified as a snake). But errors based on sensory cues are not the only things described by the users of these drugs. Anyway, are we not now being told by at least some in the scientific community, as we have been told for centuries by other communities, that the world we think we know is an illusion, indeed an error based on a whole range of sensory cues? Even if this is somewhat of an overstatement, what we are being told is that the world we perceive, if not exactly an illusion, is not even close to being all of reality, that realms of reality beyond our perceptions are available for exploration by other means. For some, these means have always included drugs, and I'm sure they still do.

William Braden, in his comprehensive survey of the psychedelic movement in the 1960s, pointed out that there are scores of psychedelic substances, both natural and synthetic, and that LSD (d-lysergic acid diethylamide)—the most popular in the West—is only one of many agents capable of producing a full-fledged psychedelic experience. Identical effects can be obtained from use of Indian hemp and its derivatives, including hashish; from the peyote cactus and its extract mescaline; from a Mexican mushroom and its laboratory counterpart, psilocybin. Hemp and peyote have been used as psychedelics for centuries, and mescaline was on the market before the turn of the twentieth century. Braden goes on to explain that LSD's uniqueness resides in the fact that it is very easy to make and megapotent.[1]

The general public knew comparatively nothing about LSD until 1963, when two professors, Timothy Leary and Richard Alpert, lost their posts at Harvard University in the wake of charges that they had involved students in reckless experiments with the drug. Leary went on to become more or less the titular leader of the popular drug movement of the '60s, which spread rapidly to campuses and cities across the United States and Canada. Leary preached the gospel of this new drug, of a new religion and a new way to find God within. Though Leary had been brought up a Roman Catholic, he was influenced by the religions of the East. His motto, and that of his many disciples, was "turn on, tune in, drop out"—meaning, turn on with LSD, tune in to the infinite wisdom in your own mind, and drop out of the meaningless status and activities of the world. Leary, and many of his fellow users, firmly believed that they were having truly religious mystical experiences under the drug. He said that he had collaborated with more than fifty scientists and scholars and that together they had arranged transcendental experiences for over one thousand persons from all walks of life.[2]

The interest generated by this research led to the formation of an informal group of ministers, theologians, and religious psychologists who met once per month. In addition to arranging for spiritually oriented psychedelic sessions and discussing prepared papers on a regular basis, this group provided the supervisory work for what was known as the Good Friday experiment. This study was the subject of the PhD dissertation of Walter N. Pahnke, a graduate student in the philosophy of religion at Harvard University. Pahnke was both an MD and a Bachelor of Divinity. He set out to determine whether the transcendent experiences reported during psychedelic sessions were similar to the mystical experiences reported by saints and famous religious mystics.

Pahnke was greatly influenced by the works of Walter Stace, whom I referenced earlier. Stace's conclusion that there are certain fundamental characteristics in the mystical experience that are universal and not restricted to any particular religion or culture (although particular cultural, historical, or religious conditions do influence both the interpretation and description of these basic phenomena) was taken as a presupposition. Pahnke decided that whether or not the mystical experience was considered religious depended upon one's own definition of religion, so he did not address this issue. He simply set out his own typology defining the universal phenomena of the mystical experience, whether considered religious or not, and then compared the mystical experience of an experimental group that had taken psilocybin with this typology. As one would expect, his typology is similar to that of Stace and the other writers referenced earlier.

Briefly, Pahnke's nine categories are as follows:

1) Unity: To his mind, this is the most important characteristic of the mystical experience, and following Stace, he says it is divided into internal and external types, which are the different ways of experiencing an undifferentiated unity.
2) Transcendence of time and space: This category refers to loss of the usual sense of time and space.
3) Deeply felt positive mood: Pahnke describes the most universal elements as joy, blessedness, and peace.

4) Sense of sacredness: He defines sacredness broadly as what a person feels to be of special value and capable of being profaned.
5) Objectivity and reality: This category has two interrelated elements. The person receives insightful knowledge or illumination on an intuitive, nonrational level and gains direct experience. The experience is considered truly real, in contrast to the feeling that the experience is a subjective delusion.
6) Paradoxicality: Accurate descriptions and even rational interpretations of the mystical experience tend to be logically contradictory when strictly analyzed.
7) Alleged Ineffability: Words fail to describe the experience adequately.
8) Transiency: This refers to duration and means the temporariness of the mystical experience in contrast to the relative permanence of the level of usual experience.
9) Persisting positive change in attitude and/or behavior.[3]

The purpose of the Good Friday study was to gather so-called empirical data about the state of consciousness experienced. It was an experiment in which psilocybin was administered to a group of volunteers in a religious context. In a private chapel on Good Friday, twenty Christian theological students, ten of whom had been given psilocybin one and one-half hours before, listened to a two-and-a-half-hour religious service that consisted of organ music, four solos, readings, prayers, and personal meditation. Data was collected during the experiment and at various times up to six months afterward.

From these data, Pahnke concluded that the subjects who received psilocybin under the conditions of this experiment experienced phenomena that were apparently indistinguishable from, if not identical to, certain categories defined by the typology of mysticism. He concluded further that the results of this experiment supported the claims made by others who have used psilocybin or similar drugs, such as LSD or mescaline, to aid in the induction of experiences similar to those experienced by mystics. His final point was that such evidence also pointed to the possible importance of biochemical changes that might occur in so-called nonartificial mystical experience, particularly the effects of ascetic practices.

The bible of the LSD cult during this decade was *The Psychedelic Reader*, which was compiled by Timothy Leary, Ralph Metzner, and Richard Alpert. The manual admits that the drug does not produce the transcendental experience. It merely acts as a chemical key. It opens the mind and frees the nervous system of its ordinary patterns and structures. In terms of what we have learned so far about reality being uncovered and explored via quantum theory, we could say that it opens the mind to other layers or realms of reality, or at least to the acknowledgment that there are such realms. The book stresses that the nature of the experience depends almost entirely on set and setting.[4] This supports the idea that, while there may be a common experience—at least a common ecstatic experience—at the base of all mysticism, it is always interpreted and understood in the light of the mystic's particular beliefs and culture.

Drawing from various sources, William Braden also constructed a typology of the central or core experiences under the drug. Again, his description sounds very much like that of the ecstatic experience of classical mysticism. Braden says that under LSD, the sense of self or personal ego is utterly lost; *I* and *me* are no more. Subject-object relationships dissolve and the world no longer ends at the fingertips. The subject sees the world as simply an extension of the body or the mind. It is fluid and shifting, and it shimmers as if it were charged with a high-voltage current. The subject has the feeling that he could melt into walls, trees, and other persons; he is keenly aware of the atomic structure of reality; he can feel the spinning motion of the electrons in his body. Braden emphasizes, however, that the subject feels that his identity is not really lost. On the contrary, he is convinced that the subject's identity is found and expanded to include all that is seen and all that is not seen.[5]

Still another writer, Sidney Cohen, reported that a change in time perception was one of the notable features that intrigued most people who took LSD. For them, time seemed to stop, or in any case, ceased to be important. They were content to exist in the moment or the here and now. Their ego boundaries tended to dissolve, and separation between the individual self and the external world became tenuous and sometimes nonexistent.[6]

These are just some of the many claims about persons reaching mystical, or at least mystical-like, experiences using a variety of psychedelic drugs. Claims such as these helped give some gloss of seriousness, even credibility, to the 1960s drug culture. To some observers and commentators, drug use was seen as a pitiful and flawed exercise, a sort of shortcut to transcendence that could in no way be compared to the classical mystic path or any serious religious commitment. To others, it was just plain delusional. However, at least on the surface, the actual ecstatic experience under psychedelic drugs does appear to have some, even many, of the qualities of the mystical ecstatic experience. Perhaps the sensory deprivation button in the brain is the locus of the cause. But so what? The brain interprets all our experiences of reality. We have learned that something being subjective doesn't mean that it is not real. Indeed, it may be the only real.

It is useful to emphasize, even while maintaining a skeptical mind-set with all of this '60s hype, that many religious traditions have included the use of drugs to achieve what they believe to be a connection—either internal or external—with the beyond. There also appears to be a common experience in the ecstatic mystical state, whether it is experienced by someone within an Eastern, a Western, a Native, or a secular humanist tradition and however it is induced. The feeling of oneness appears to be the constant. The difference lies in how it is interpreted and understood and how this experience affects the individual.

It is very easy, even natural, for those of us raised in the secular, science-oriented, Western culture to be more than skeptical toward any or all of these essentially emotional avenues to an appreciation of reality and truth, especially with the use of drugs. The seemingly goofy psychedelic movement of the '60s embedded this essentially cynical outlook. However, perhaps a better way to look at it, in retrospect and in light of new understanding from quantum theory, is that it was simply a rather typical, essentially easygoing, Western manifestation of something much deeper—something not just Eastern but Western as well. That is, a universal desire to be in touch, however imperfectly, with something beyond, something more than the immediate reality that we experience every day, to know, verily know, that there is more than merely me or us.

Clearly, this is the goal of all mysticism, whether drug-induced or not, as well as of all religion, philosophy, and science (or natural philosophy). Mystical apprehensions of otherness and beyondness—if we use the broader definitions from von Hugel and others referenced earlier rather than the more limited ecstatic experiences, while in no way eliminating or denigrating these ecstatic experiences—are part of the intellectual and emotional landscape of most, perhaps all, great thinkers and perhaps of all of us.

As I have said, the new insights of relativity and quantum mechanics have, most importantly, made such discussion and debate respectable again in the West. This is no small thing, since we all know that political correctness and current truths can and do swamp intelligent discussion and debate. A true waltz in wonder requires an openness to wonderful and extraordinary possibilities, both intellectual and emotional—indeed, an openness to goofyness.

The doubt and uncertainty now at the basis of science may allow new opportunities for this to happen. However, it is important to remember that the uncertainty in science does not indicate less rigorous thinking, nor does it in any way signify a turn to the purely instinctive and emotional. That is why its method, its openness, and its encouragement of discussion, debate, and skepticism must continue to be enshrined in our psyche.

Chapter Twelve
The Origins of Disinterested Reason

While there is much speculative ground to be tilled in linking the insights of quantum physics with those of the mystic traditions and Eastern religions, the ground is equally fertile in the West. As I have indicated on a couple of occasions, the questions about reality and truth now occupying scientists have occupied philosophers throughout the history of Western civilization. Indeed, they are universal questions that have occupied the minds of thinking people probably since people could think.

Natural philosophy, which became known as science only recently, has always rigorously and rationally pursued the answers to these questions. However, after Newton, with mathematics as a primary means and particularly as experimental validation became a necessity, the purview of natural philosophy, by necessity, narrowed. This severely truncated view underrated the breadth of philosophical and metaphysical speculation on which Western thought is based. What is exciting about the new scientific perspectives introduced by relativity and quantum mechanics, but especially by the latter, is that the wider metaphysical speculations are now being reintroduced.

The term *rational* will be used here in its generally defined sense as being based on or derived from reasoning and not just in the more restricted sense of reasoning from observable facts but of reasoning period. Its opposite is *irrational*, which is based on emotion rather than intellect and is often connected with mysticism. The derivative term, *rationalism*, is defined as the practice of accepting reason and the intellect as the only source of true knowledge. Essentially, it is the faith that reason, and reason alone, can discover truth. This faith appears to be a gift from the Greeks and peculiar to the West. While it does not decry intuitive or mystical insights, indeed even values these gifts from unknown sources, everything must pass the test of reason before it can be accepted as real knowledge or truth.

Both philosophy and science are based on continuous rational questioning, within the context of a faith that believes the questioning has meaning and purpose. In the West, we generally understand that this frame of mind came into being in earnest with the Greeks, probably starting in the sixth century BC. Greek philosophy, natural and otherwise, is certainly the gold standard for Western civilization. One reason, perhaps, that this sort of intellectual speculation was able to evolve in Greece rather than in some other of the older, vibrant civilizations was that there was no overpowering religion with sacred texts to impede such speculations. Whatever one might think of the value of organized religion, it is generally not conducive to intellectual adventure.

The great Canadian scholar Harold Innis, in his monumental study of the history and effects of communications, said that Greek culture was particularly creative because it was able to escape the stupefying tendencies of writing until a comparatively late date. When the tradition was finally committed to writing on papyrus scrolls, the flexible phonetic alphabet more accurately reflected its creative power than would have been possible earlier.

Innis was a great believer in the creative effects of oral communication, not only for the Greek civilization but also for that of the Hebrews and the Christians. But unlike the latter two, "the Greeks had no bible with a sacred literature attempting to give reasons and coherence to the scheme of things, making dogmatic assertions and strangling science in infancy."[1] The relative absence of dogma allowed for the evolution of truth. However, Innis believed that the stultifying effects of writing, even with the flexible phonetic alphabet, could be seen in Greece by the second half of the fifth century BC. As early as the fourth century BC, Plato recognized the danger inherent in writing and attempted to save the remnants of Greek culture by reflecting the oral tradition in the style of the Socratic dialogues. Socrates knew that "the letter is designed to kill much (though not all) of the life that the spirit has given."[2]

In Eastern civilizations, the emphasis tended to

be placed on the mystical and emotional rather than the rational. Perhaps what saved the Greeks from this emphasis was the rise of the schools of natural philosophy in Ionia.

The first to develop was in Miletus on the Ionian coast. Thales of Miletus speculated that all things are made of water—that is, that the most fundamental element of existence was water. Anaximander, also a Milesian philosopher who lived a bit later than Thales, questioned the assumptions of his predecessor and said that the primary stuff of nature could not be the same as one of its special forms, that it must therefore be more fundamental than water. He called it the boundless, infinite material everywhere. The next famous Milesian thinker was Anaximenes. He believed that the basic substance of the universe was air. This is what the soul is made of, and it keeps us and the world alive. These speculations, a passionate belief in the pursuit, and yet a disinterested (in the sense of not ideologically or religiously driven) search for the basic element of reality are very much the same as current scientific theorizing—indeed, of scientific theorizing since that time.

The Milesian school dealt with intrinsically practical matters. Not so the Pythagorean movement that emerged off the coast of Ionia on the island of Samos. Members were part of a religious community founded by Pythagoras. However, in spite of this, there was a strong scientific and especially mathematical side to Pythagorean philosophy. The Pythagoreans were impressed by the fundamental importance of numbers in the world. They saw that everything is numerable and that the relation between two related things may be expressed according to numerical proportion. Their belief in the intrinsic nature of mathematics—that math principles were the principles of all things, the basis of all reality—was clearly metaphysical and reflected their mystical and religious origins.

Bertrand Russell says that the Pythagorean focus on mathematics also gave rise to the later theory of ideas or universals in the works of Plato. It is this metaphysical belief that has dominated math and science to the present day. All the currently proposed truths of quantum physics are mathematical truths, no more, no less.

Another Ionian, Heraclitus of Ephesus, lived around the turn of the sixth century BC. He said that the real world consists of a balanced adjustment of opposing tendencies. Behind the strife of opposites lies a hidden harmony. But the ongoing strife is the creative principle and keeps the world alive. This view required a new kind of fundamental matter that would emphasize activity. Heraclitus chose fire, probably using it metaphorically in that it dramatically illustrates his theory that nothing stays still, that everything that happens is a process of exchanges. "All things are in flux," he is supposed to have said. Nothing is stable, nothing abides. Yet, paradoxically, perpetual change keeps things as they are. There is unity in diversity and difference in unity. For Heraclitus, the conflict of opposites, rather than undermining the unity of the One, is essential to it. In fact, the One only exists in the tension of opposites.

Reality for Heraclitus is One, but it is Many at the same time.[3]

Here again is a classic issue that dominated Greek and Western philosophy since the time of the Greeks and still dominates theoretical physics: the problem or the question of the One and the Many, the relation between them, and the character of both. The One is the underlying and fundamental thing or principle or equation for reality that philosophers and scientists have been searching for, whether it is earth, air, fire, water, or atoms, the universal spirit of the mystics, and/or the theory of everything of quantum physicists. The Many are the many sensory and experiential manifestations of reality that we see around us and experience daily. How does the One relate to the Many and vice versa?

In the Greek and Western traditions, this question is pursued rationally. In the East, it is pursued mystically. The issue arises because the Many are constantly given in experience as the result of sense impressions. But mystics, philosophers, and theoretical physicists all believe there is a unity—a One—behind the world of experience. They try to arrive at a comprehensive view of reality, which sees the Many in light of the One or reduces the Many to the One. We have seen this in the mystic's vision of an undifferentiated unity of all things, which is an essentially emotional understanding, and in the physicist's attempts to formulate a theory of everything, which is an essentially mathematical rational understanding.

The issue developed further in the philosophy of

Parmenides, who lived during the first half of the fifth century BC. Sounding very much like one side of a classic debate in theoretical physics between the idealists and the materialists, he asserted, that reality was material, that the basic stuff of the universe is matter. He was, therefore, a true materialist. Matter comes first. Our seeing something doesn't make it happen. Parmenides believed that basic matter is uncreated and eternal. The world is full of matter, and empty space does not exist. Too bad if it is counter to our sense experiences. He believed that we must write off our sense experiences as illusory.

A criticism of Parmenides brought a new approach to the question of reality and the questions of the One and of the Many. This was provided by Empedocles, a citizen of Akragas in Sicily. Parmenides had said that matter is without beginning and without end; it is eternal. Empedocles accepted this. However, it is obvious that change happens and is inevitable, so he had to reconcile this with the theory of Parmenides. Empedocles did this by proposing that an object as a whole begins to be and ceases to be, as experience shows it does, but it is composed of material particles, which are themselves indestructible. These are the fundamental and eternal kinds of matter or elements—earth, air, fire, and water. He called these the "roots of things." They were later called the four basic elements by Aristotle, and as such, this famous theory dominated chemical science for centuries.

However, even with these elements, Empedocles was

faced with the question of how, from these elemental objects, the world of our experience comes into being. What force is responsible for the cyclical process of nature? To explain this, Empedocles postulated two active fundamental forces: Love and Hate or Harmony and Discord. In spite of their names, these forces were conceived of as physical and material forces—Love or Harmony bringing the particles of the four elements together and building up; Hate or Discord separating the particles and destroying the objects.[4]

Further and deeper thought was given to the question of what constitutes ultimate reality and how to reconcile the One with the Many by Anaxagoras. Although originally from Ionia, he was the first of the philosophers to settle and live in Athens. Like Empedocles, he accepted that fundamental matter neither comes into being nor passes away but is unchangeable. But he did not agree that the ultimate units of reality corresponded to the four elements—earth, air, fire, and water. He believed that anything that has parts qualitatively the same as the whole is itself ultimate and cannot be divided. So he believed that the four elements must be composed of many qualitatively different particles. This assumption that matter is infinitely divisible sounds very modern, and this is the first time it was proposed. It sounds very much like an early precursor of "sizeless points" or "strings with no mass."

Sounding much like the Nobel physicist Robert Laughlin to my mind, Anaxagoras said that the objects of experience arise when these ultimate particles are brought together in such a way that particles of a certain kind predominate in the resulting object. While Laughlin does not speculate as to exactly how or why this coming together in specific ways happens, Anaxagoras did. In so doing he made a unique contribution to philosophy, he introduced an active, creative principle that he called Nous or Mind.

This takes the place of the active principles of Love and Hate in the philosophy of his predecessor Empedocles. Anaxagoras believed that this Nous had power over all things that have life and that it set in order all things that were to be, all things that were and are now, and all things that will be. He did not believe that Nous created matter, since matter is eternal, but its function seemed to be to organize matter to form the reality we know and experience. While he appeared to be somewhat uncertain as to whether his Nous was itself material or spiritual or both, his major contribution is that he first introduced this intellectual and spiritual principle that is still the basis of uncertainty, discussion, and debate today: the so-called God particle.

The atomists borrowed from all previous theories, especially those of Parmenides and Heraclitus. From Parmenides, they borrowed immutable elementary particles, while from Heraclitus came the notion of ceaseless movement. The founder of the Atomist School is generally considered to be Leucippus, who, like the earliest philosophers, came from Miletus. Much was added to the theory by later contributions from Democritus of Abdera around 420 BC. According to him, there are an infinite number of indivisible units called atoms, which are much too small to be perceived by the senses. Atoms are infinite in number, are all supposed to be the same in composition but differ in shape. What distinguished atoms was that they could not be divided or broken up in any way. The definition of *atom* is that which cannot be cut. However, in moving about, collisions or "comings together" occur. Out of these collisions come the four elements—earth, air, fire, and water—and the world is formed.

Importantly for the atomists and for the future development of science, no external power or moving force is assumed as a necessary cause for this primal motion. Neither the Love and Hate postulated by Empedocles nor the Nous of Anaxagoras is postulated. The atomists took the position that atoms simply existed in the void in the beginning, period. From that beginning alone arises the world of our experience. The influence of this notion on the development of Western science is obvious.

The belief in these fundamental and eternal material elements has persisted throughout the history of science and is the basis of current particle physics. When physicists finally proved through laboratory experiments that atoms do exist, they shortly thereafter proved that they were not indivisible. On the contrary, ever smaller components of atoms were discovered until today physicists speculate an infinite number of ever smaller and smaller elementary particles. But even when quantum theory makes these ever smaller particles unlikely, they are still described as particles.

At the earliest stages of Greek philosophy, there is the notion of unity, the notion that things change into one another and therefore there must be some common basic element of nature, whether that be water, air, fire, earth, or some combination of these. But they all believed—and believed is the right word—in one ultimate principle. For this was very much a faith position. It still is. It was and is beyond any proof, at least any rational proof.

The central question for the pre-Socratic Greek philosophers centered on the external world, just as Newtonian science did. They were interested mainly in the Object—the ultimate physical principle of the world. The Self or Man, while not excluded, was secondary. Their principal question was "of what is the world composed?" The next period moved way beyond this primarily material focus, to concentrate more on the subject, as did the movement from Newtonian to quantum physics.

The three greatest figures in Greek philosophy—Socrates, Plato, and Aristotle—were all associated with Athens. The first two were born there, and Aristotle studied and taught there. We know very little about the life of Socrates and rely on the writings of his pupil, Plato, for information about him. Though Plato tells us that Socrates insisted that he knew nothing, Socrates believed firmly that knowledge is not beyond our intellectual reach—that is, beyond the reach of our reason.

What mattered to him was precisely that we should constantly pursue knowledge, that we should pursue truth, and that we should pursue it rationally. He believed that the cause of evil is ignorance. Sin is a lack of knowledge. If we could only know—verily know—we would not sin. Therefore, to reach the Good, however defined, we must have knowledge. This link between good and knowledge is a heritage from Socrates through Plato and Plato's student, Aristotle, to Western philosophy and science.

When Socrates says he knows nothing, he is speaking ironically, although he is also making a very serious point. In fact, of course, he knows a great deal. He was thought of as the wisest man in the land. However, what he knew very clearly—what has not been as successfully passed down through philosophy and science in the West but is worthy of universal and current understanding—is that what he knows and what we all know is as nothing when set against the infinite vastness of what we do not know. We are all dancing in the dark. When any one of us truly understands this, we must say that we know nothing. That is indeed wisdom. A little of this humility can be seen in the early quantum physicists.

Plato was born in Athens in 428 BC. He has probably had a greater influence on philosophy in the West than any other person. It has been said that all subsequent philosophic thought in the West is but a footnote to Plato and that Plato's philosophic thought is not *a* metaphysic but the one and only metaphysic. He was the philosophic heir of Socrates and the pre-Socratics.

He was the founder of the Academy, which has been called the first European university, and the teacher of Aristotle, who was one of the Academy's earliest students. Studies in the Academy were similar to those of the Pythagorean schools, with an especially strong emphasis on mathematics but also studies in astronomy and the physical sciences. Plato believed that the best training for a practical public career was the pursuit of science—disinterested science—for its own sake. The main objective of the school was to move the thinking processes of its students from the ever-changing world of experience to the unchanging framework lying behind it—from the specific to the universal, from the Many to the One.

This sounds very modern, and it is. The Platonic focus, to find the unchanging behind the world of sense perception—what reality or truth is—is the prime objective of theoretical physics. The process used was inherited from Plato and the ancient Greek philosophers. Much as quantum physicists do, Plato took the position that true knowledge learned via the senses is not possible. Like them, he believed that one must come to know what is universal and abiding in order to discover true knowledge or truth; indeed, that truth *is* universal and abiding.

For him, and for Western science, the way one comes to know this is through disinterested reason, which is reason for its own sake rather than reason in the service of some ideology or theology. Reason must be supported by argument, which is why Plato used the dialogue method in his writings. In the absence of argument, there is no knowledge. This and the importance of mathematics in the disinterested pursuit have remained of central importance throughout the history of science in the West. Indeed, they are basic to the scientific method.

For Plato, argument was basic to the philosophic method, through which—as he explains with his famous Allegory of the Cave—one is able to see beyond the shadows of our day-to-day experience to the "sunlight of reason and truth," to what is real and abiding. The philosopher wants to know, for example, the essence of beauty that lies behind beautiful things or the essence of goodness that lies behind good things. Much like the theoretical physicist's belief in the reality of mathematical forms, Plato believed that the conceptual ideas of absolute beauty and absolute goodness were real and exist outside of and somehow beyond our human ability to conceive of them. He called these Ideas or Forms. They are not merely products of our human minds. We conceive of them because they exist, because they are real and we see them illuminated, however indistinctly, in beautiful and good things that we see around us.

As Bertrand Russell and others have said, these Ideas or Forms are mathematical in nature. Some contemporary theoretical physicists and mathematicians, notably Roger Penrose, describe mathematical truths in this Platonic sense. "There is something absolute and 'God-given' about mathematical truth. This is what mathematical Platonism is about. Any particular formal system has a provisional and 'man-made' quality to it.... Real mathematical truth goes beyond mere man-made constructions."[5]

For the philosopher, the mind moves metaphorically upward from the many specific examples of things in the world to their conceptual Ideas or Forms and upward to the highest concept that is the Good. The Good is inferred to be "the universal author of all things beautiful and right, and the source of truth and reason."[6] It is also described as the One. It is said that all the various Forms or Ideas and the many specific examples of each that flow from them owe their being in some way to this One. This is certainly analogous to the faith of modern theoretical physics that through disinterested reason and knowledge of mathematical forms, which physicists believe are real and not mere figments of the imagination, the mind moves ever upward toward the understanding of reality and the One master mathematical formula.

A philosopher is literally a "lover of wisdom." But not just anyone who might be curious for knowledge can be defined as a philosopher. Rather, a true philosopher is one who is caught up in what may be described as a vision of truth, one who believes passionately that there is such a thing and longs to find it and define it. This describes the best of the theoretical physicists. That is what the theory of everything will be: a mathematical vision of truth, an expression of the unchanging reality. It, as with other mathematical visions such as strings and branes, is passionately sought by many physicists.

They are the inheritors of the legacy of the Greek philosophers, and Greek philosophy is exemplified in the Platonic dialogues. Physicists understand, as Plato understood, that the process of moving from the world of shadows to the world of sunlight is one of intense mental discipline and education. This does not refer to ecstatic enlightenment following either physical deprivation or the ingestion of drugs. This requires long and hard work and a deep understanding of mathematics. Enlightenment comes only to those who have prepared themselves through years of work and only in those areas for which they have prepared themselves.

To draw one final connection between Plato and contemporary quantum physicists, just as Plato broke away from the de facto materialism of the pre-Socratics, asserting the existence of immaterial and invisible reality that is far deeper and more real than the material world, quantum physicists broke away from the materialism of Newtonian science. While agreeing with Heraclitus that the sensible world is in a state of flux, or a state of becoming, so that it can never be said to actually be, Plato asserted that behind this state of flux there is a stable and abiding reality that can be known and that is, in fact, the supreme object of knowledge. This reality has mind and life and—yes—soul. So there is spiritual movement in the real. This sounds remarkably like at least some of the interpretations of reality proposed by quantum theorists.

While Plato was still teaching, Aristotle was born in about 384 BC at Stagira in Thrace. At the age of eighteen, he came to Athens to study under Plato at the Academy. He remained a member of the Academy for over twenty years until the death of Plato in 348 or 347 BC, and he was the first great critic of Plato. While Aristotle's writings on subjects such as logic, ethics, and nature have all been influential, his work on metaphysics is especially relevant for our purposes. Bertrand Russell explains that what we now call metaphysics did not go by that name in Aristotle's time. *Metaphysics* simply means "after physics." Aristotle's book gained this title because an early editor put it after *Physics* in the arrangement of his works. It would have been more appropriate to place it before physics, since Aristotle would have called it "first philosophy," meaning a discussion of the general preconditions of inquiry. However, the name *metaphysics* has gained currency.[7]

Metaphysics begins with the statement that "all men desire to know." However, there are different degrees of knowledge. Aristotle places the person who seeks knowledge for its own sake above the person who seeks knowledge of a particular kind with an aim in mind of attaining some practical effect. The science that is desirable for its own sake is the science of first principles or first causes—a science that arises in wonder. This is a perspective that has come down via universities in the West through the ages and is exactly the perspective of modern quantum physics, notwithstanding the practical spin-off results of the theory.

Thus, metaphysics, for Aristotle, is wisdom par excellence, and the philosopher or lover of wisdom desires knowledge about the ultimate cause and nature of reality and desires that knowledge for its own sake. Wisdom deals with the first principles and causes of things; therefore, it is universal knowledge of the highest order. This means that it is the science that is furthest removed from the senses, the most abstract science, and so is the most difficult and involves the greatest effort of thought.[8] Where could one find a better description of quantum physics and many, even perhaps most, quantum physicists?

Aristotle attempted to replace Plato's theory of Ideas or Forms with a new theory of his own. Essentially, his critique of Plato is the classic down-up versus up-down theory. Plato's theory was up-down. He believed that the highest principle or universal, the Good, comes first, and from it comes the other universals (e.g., love, desire, kindness, but also man, horse, dog, house, etc.), and from these come specific material examples or instances of those universals. The universal Ideas that are highest come first, and from them comes the sensory world that we know. Aristotle's theory, by contrast, was down-up. Aristotle believed that we gain our appreciation of universals from the specific examples we see around us. From seeing specific horses, for example, we get a conceptual framework for the generic horse.

These different perspectives are both historic and current. They are incorporated, for example, in the theory of evolution versus the theory of creationism. They are at the base of one of the most effective critiques of Christianity, *The Essence of Christianity*, written in the middle of the nineteenth century by Ludwig von Feurbach, who expounds the theory that God is made in the image of man rather than the traditional Judeo-Christian theology that man is made in the image of God. However, it is quite possible for a thinking person to understand and accept both perspectives. One can understand that any image that a human being might have of God would have to be a humanlike image, while at the same time believing that God is beyond all images. It is also possible to believe that evolution is the only logical and consistent theory about our world's beginning and development while believing in the necessity of a purpose and direction beyond the process itself. Both are based on one's faith perspective. We may be dancing in the dark, but we can dance in faith as well as wonder.

Both Aristotle and Plato so danced, as did their philosophical predecessors and successors. The statement by Einstein that "the most beautiful experience we can have is the mysterious" is in the spirit of this tradition. But it was also Aristotle's and Plato's belief in and practice of reasoned inquiry that has shaped the West. Unlike mysticism, the truths arrived at by this approach are always open to further inquiry and further research. They do not claim to be ineffable. Nor do they take refuge in silence and the inability to communicate. The philosophic tradition of ancient Greece, exemplified in the works of Plato and Aristotle, was thus one of enlightenment and liberation. It aimed to free the mind, to remove the fear of the unknown by demonstrating that the world is accessible to reason. The philosophic tradition aspired to the pursuit of knowledge and truth. Disinterested inquiry was itself regarded as good. It was through this, rather than through religious mysteries, that one could achieve the good life.

This perspective was due at least in part to the influence of mathematics, which was as fundamental to Greek natural philosophy as it is to modern science. Bertrand Russell, himself a mathematician, explains why mathematics is so important. In the first place, while math problems are not always easy to solve, they are clear and simple in comparison to other types of problems. In the second place, there is a well-established and clearly understood procedure in their demonstration. Finally, mathematical conclusions, once understood, cannot be doubted.

The reason for this third point is that, unlike other arguments, part of the procedure of the mathematical argument is that the premises are understood and accepted by everyone, whereas this is rarely the case in other types of arguments. In mathematics, there are no facts or truths that might cloud the mind or call for comparisons outside the mathematical structure under consideration.[9] Because of this certainty, philosophers in the West—certainly natural philosophers both ancient and modern—have argued that mathematics provides knowledge that is superior and more reliable than that from other fields of inquiry. Some have even said that only mathematics provides real knowledge and the only truths are mathematical truths. Certainly quantum theorists would agree with this.

Be that as it may, besides the simplicity of its problems and the clarity of its structure, mathematics provides plenty of scope for the creation of the beautiful. As we have seen, today's theoretical physicists, like their ancient Greek ancestors, are more than aware—I would say emotionally entranced—by the beauty of the math, the beauty of the equations. A theory of everything would be beautiful, elegant, and economic. As Russell explains further:

> The Greeks, indeed, possessed a very acute sense of aesthetics.... The sentiment expressed by Keats in saying that truth is beauty is a thoroughly Greek conception. It is precisely the kind of thing a Platonist may well feel in contemplating the geometrical proportions of a Grecian urn. The same holds of the structure of mathematical proof itself. Notions like elegance and economy in this field are aesthetic in character.[10]

The Keats poem, referenced by Russell, is featured on the introductory page, unnumbered, of Ian Stewart's book, *Why Beauty Is Truth*:

> When old age shall this generation waste,
> Thou shalt remain, in midst of other woe
> Than ours, a friend to man, to whom thou say'st,
> "Beauty is truth, truth beauty,"—that is all

Ye know on earth, and all ye need to know.

—John Keats,

"Ode on a Grecian Urn"[11]

Finally, before we leave the ancient philosophers, I want to touch again on the theories of Plotinus and Neoplatonism. These theories are referenced in important ways by Shimon Malin, one of the scientists who has a great deal to say about the quantum revolution. Malin believes that the theories of Plotinus have many similarities and bring much enlightenment to quantum mechanics. Plotinus' theories also end the ancient phase and act as a sort of bridge to the Middle Ages in the West. Neoplatonism completed a transformation of classic Greek philosophy, specifically the philosophy of Plato, into mysticism and religion from a disinterested rational and intellectual pursuit of the mysterious to a more mystical pursuit of such a goal. It influenced the development of Christianity and helped establish the character of metaphysical inquiry up to the time of the Enlightenment.

Plotinus was the greatest of the Neoplatonist scholars and lived much later than the philosophers so far referenced (AD 204–270). He was not Greek but was born in Egypt and finally settled in Rome. He was much influenced by mysticism, and this influence shows clearly in his writings. The focus of Plotinus' philosophy is his theory of three ultimate levels of reality, a sort of trinity in descending order from the basis of emanations from the highest order. He called the three levels the One, Nous, and Soul, in that order of priority and dependence.

The One of Plotinus is the One of the mystics, about which nothing can be said other than "it is." He sometimes speaks of it as God and sometimes, in the manner of Plato, as the Good. However, the One is beyond all thought and all being, ineffable and incomprehensible. This does not mean that the One is nonexistent. On the contrary, it simply means, in the language of the mystics, that the One transcends all beings with which we have experience. How then do we experience it? Plotinus used the metaphor of emanation to explain this. This metaphor rejects both the theistic and the pantheistic theories. Plotinus' God—the One—is totally separate from all else and not involved in creation in any way.

The next element, Nous, simply emanates, or comes forth in some mysterious way, from the One. Nous may best be described as thought or mind or perhaps self-consciousness or intuition, although an exact description is probably not possible—certainly not by this writer. The Nous has a twofold object: the One and itself. The Ideas or Forms of Plato exist in the Nous. By exercising our own minds in the direction away from sense, we can come to know the Ideas and then the Nous and through it the One, of which the Nous is a sort of image.

The third member of this so-called trinity and the second emanation—this time from the Nous—is called the World-Soul. This World-Soul is incorporeal and indivisible, but it forms the connecting link between the supersensual world of the Nous downward to the world of sense and nature. Individual souls emanate from this World-Soul. Finally, below the sphere of Soul is the material world.

Regarding the individual soul, in its highest part, it is uncontaminated by matter and remains firmly connected to the intelligible world. When it enters real union with the body, it is contaminated by matter, and there is the necessity of an ethical ascent, with the objective of union with the One. In this ascent, the soul has to rise above sense perception. It has to turn toward Nous and focus on philosophy and science.

The process of rising above sense perception to the pursuit of reality via philosophy and science reflects both Plato and Aristotle, as it does, in its own way, current theoretical physics. But then Plotinus leaves the rational pursuit that is central to the Greeks and to modern science and moves to classic mysticism. Following the soul's focus on philosophy and science, a higher stage carries it beyond discursive thought to union with the Nous. All these stages are, in fact, a preparation for the final stage: classical mystical union with the One or God.

So Neoplatonism passed from philosophy into mysticism. This is, of course, a slippery slope for all metaphysical speculation, both ancient and modern, especially for speculation based on the philosophy of idealism instead of materialism (i.e., the theory that reality is spirit or thought or ideas instead of matter or particles). Quantum theory is very much idealist rather than materialist.

Chapter Thirteen

St. Augustine and the Rise of Interested Reason

Speculative philosophical thought, beginning with the Ionian Greeks and utilizing so-called disinterested reason, focused first on cosmology to account for the phenomena of nature. The early natural philosophers looked for the one element that was basic or fundamental to the natural world. While skeptics had an important role to play during this period, those who believed in the reality of one common element at the base of the many appearances fell roughly into two camps. One camp pictured the cosmos in terms of form, and the other pictured the cosmos in terms of matter. This became the historical distinction between the idealists and the materialists. Their different perspectives proved to be irreconcilable since, while each was eminently reasonable, each was also completely arbitrary and based on faith.

Philosophy and science developed in ancient Greece as an attempt to determine reality and the frontiers of nature by representing nature as a closed system of orderly relationships, much like it is still viewed today. It then proceeded to take apart and study the elements. Its objective was to uncover the basic, principal underlying phenomena, that which is ultimate being or reality. Whether matter or form, this element was recognized as the first creator or substantial cause.

But once this fundamental principle was recognized, the question of how it related to the everyday world of sensory experience—how the One related to the Many—presented itself. So a second fundamental principle evolved in order to relate this world of pure being to the world around us of becoming. The Greeks called this the principle of movement. Finally, a third principle was developed to illustrate the connection between the first two and was described as the principle of intelligibility or order. So, via a purely rational approach, the ancient Greek philosophers developed three independent first principles to account for being, movement, and the relationship between them.[1]

Plato dismissed these first principles as inadequate, as mere theories or opinion. For him, none of them touched on reality or truth. No necessary relationship had been established between the principle of being and that of movement. The classic problem of the One and the Many remained. He believed that the problem lay in a lack of understanding the third principle of intelligibility or order. So his philosophy set out to elucidate this in his concept of the Ideas or Forms, demonstrating how, via the method of rational dialectic, the mind could be elevated to these pure Ideas and through them, finally, to the one fundamental Idea—the one first principle—which he called the Good.

St. Augustine believed that Plato's Good was God. He regarded Plato's philosophy as far superior to any other; in fact, he himself is often referred to as a Neoplatonist. His criticism of Plato came because, although Plato had discovered God as the prime cause, this God was purely transcendental, and therefore Plato was unable to display God's immanence while explaining the reality of the material world. Therefore, as the materialism of the natural philosophers failed to take sufficient account of the problem of mind, so this idealism of Plato failed to take sufficient account of the problem of matter.[2] The problem of the One and the Many remained.

The pursuit of the solution to this problem, and the use of reason in its pursuit, took a distinctly different turn with Christianity and the greatest "father" of the Christian Church: St. Augustine. Christianity came into existence as a revealed religion, given to the world as a doctrine of redemption and salvation and love, not as an abstract theoretical system. It dismissed the Greek and Platonic belief that reality could be perceived by the disciplined mind alone as illusion and thus rejected the dictatorship of intelligence, or the religion of science. Against this, it argued that the true starting point for thought and action must be found in revelation. Therefore, contrary to the faith of Plato and the Greek philosophers, the question of prime importance was not just the quality and capacity of thought but the presuppositions, or the faith, that govern this thought.

Faith in the God of revelation was proposed to be indispensable to full understanding. This, Christianity believed, provided a principle through which it was possible to develop a fully adequate picture of nature and the physical world, distinctly different from that of the classical world. In this picture, the cosmos is a world of concrete substances—real matter—much like it was conceived to be centuries later by Newton. For the early Church and for St. Augustine, the world of nature was neither self-generating nor self-fulfilling but totally dependent on the intelligent and beneficent support of God as its creator and preserver. The concept of the world being created ex nihilo (from nothing), indicated this sense of direct and immediate divine activity. Here, the One (God) would be best described as form, whereas the world and the cosmos are clearly matter.

St. Augustine was born in AD 354. He entered upon his life work as Bishop of Hippo in the twilight of the Western Roman Empire in AD 395. He therefore lived and worked in the fourth and fifth centuries AD. Although moral and intellectual foundations of the world around him appeared to be shattering and there was a revolt from the spirit of Greco-Roman philosophy and science, he refused to believe that all that had gone before was futile, that the secular efforts of mankind had been in vain. He refused to give up on reason or to turn from honest skepticism to some kind of primitive faith sustained by an arbitrary will to believe.

In the concept of the Trinity, which appears different in essence from but reflective of the three first principles of ancient Greek philosophy, St. Augustine found what he believed to be a concept capable of saving both the reason and the will, as well as a solution to the problem of the One and the Many. It saved the reason because, while denying its classical pretensions to discover ultimate truth, this new concept nevertheless affirmed the existence of an order of truth and value that, being a part of the world as well as beyond it, was possible for one to apprehend. In saving the reason, it also saved the will, because it imparted to the will that element of rationality without which the will would become mere subjective willfulness.

The Trinity apprehended the creative or first principle—God, the One—as a single essence that is fully expressed in three ways: the Father uncreated, the Son uncreated, and the Spirit uncreated. For St. Augustine and for the Church, this concept bridged the gap between the sensory world of experience, the Many, and the primary or first cause that is beyond the senses, the One. In this formula, the first expression—God the Father—as the creative principle is, strictly speaking, unknown and unknowable, except as it—or he—manifests in the second and third. The second expression—God the Son—is the principle of intelligence, the *logos*, literally defined as "word" or, sometimes in Greek philosophy, as reason itself manifested by speech. It reveals itself as the order of the universe and was personified in Christ. The third expression—God the Spirit—is the principle of motion in the universe.[3]

St. Augustine believed that this interpretation of the

original revelation personalized and humanized what had previously been, in the Greek concept of the three first principles, a strictly intellectual, reasoned exercise. He believed that the revelation of God in Christ revealed what classical science had always sought: the connection between the One and the Many (i.e., between the first or fundamental principle, that of pure being, and the second principle, that of movement or becoming; the world of sensory perception that surrounds us). For him and for the Church, Christ was that connecting principle—what for the Greeks was the principle of intelligibility or order.

Here then was a clear metaphysic of ordered progress; it was a reasoned argument that was finally beyond reason, beyond physics and science. In so positing, St. Augustine believed, the principle of intelligibility did justice to the element of truth contained by both classical materialism and classical idealism, while avoiding the errors and absurdities of both.[4]

Now whatever else this concept of the Trinity may be, it is certainly a most complex intellectual construct, just as complex and just as reflective of reason as the ancient Greek concept of the three first principles. Far be it from me to attempt further elucidation. Suffice it to show that St. Augustine and the other Church fathers were working within the structures of their reason just as were the ancient Greek philosophers. The only difference, admittedly a key difference, was their faith that they were also working within and on the basis of a truly original revelation.

For them the doctrine of the Trinity was a means of emancipation—in effect, of salvation—from the ignorance and blindness that results from a misunderstanding of the possibilities contained in the instrument of our apprehension and from its resulting misuse. When we receive this emancipation, we understand that, whether reality is understood as form or matter, we are simply worshiping abstractions of our own fancy, gathered from our world of sensory experience. To achieve this emancipation we are not required to forego reason but rather to open our eyes to the existence and activity of the *logos*, the creative principle, within ourselves. This is to discover and more fully understand our consciousness.

St. Augustine understood that science, as a capacity for ordered knowledge, involves mental processes that are distinctly human. These processes presuppose the existence of a mind that can reason. However, this ability to reason alone provides no guarantee of the ability to discover or uncover truth. In other words, it provides no title to divinity. Rather, it merely represents the fulfillment of one's nature as a human being. This peculiar human capability remains dependent on the creative principle. St. Augustine believed that mankind alone has this principle because we are made in the image of God. Thus, for him, this creative principle is the Christian principle, manifested perfectly in Jesus Christ.

A similar continuity is exhibited in the relationship between the knowledge of the "exterior" and that of the "interior" man (i.e., between the awareness of objects and the awareness of being aware). It is precisely this awareness that gives human beings their character and enables them to fulfill their destinies as human beings. Just as the manifestation of God in the human form of Jesus Christ was the connecting link between being and becoming, between the One and the Many, so human consciousness, in faith, provides that link.

Leaving aside St. Augustine's and Christianity's theological understanding of this perspective for the moment, it sounds very modern. Essentially, as I understand it, this perspective puts human beings (i.e., consciousness and the nature of consciousness, *logos*, or indeed, reason) at the center, as the basic premise of all philosophical, scientific, or theological speculation—of any and all concepts of reality or truth. As such, it connects to quantum theory and to a primary critique of science since Newton, which is that the major shortcoming of Western science has been that it ignores the nature of human consciousness.

St. Augustine makes the point that has been made most strongly in the last century by quantum physicists that the person cannot be separated from nature, the experimenter cannot be separated from the experiment, and that so-called third-person objectivity in science is a myth. He saw clearly that this traditional classical scientific perspective, which regardless of his warning came back fully formed during the scientific Enlightenment and remained central to scientific truth almost to the present day, can lead one to a fatal misapprehension: the belief that there is evidence in one's consciousness of existence and activity that one embodies some sort of divine essence that lifts one above the natural order of which we are a part and allows us to examine it objectively.

For St. Augustine, mankind's indulgence in such fancies is the cause of sin and error from which, as long as these theories are entertained, there is no possibility of escape. The alternative is to recognize oneself as created, with one's consciousness of self in some mysterious way dependent on an outside source of being, wisdom, and power in whose image one is made. This must be accepted on faith—a leap of faith to use the words of the Danish philosopher Soren Kierkegaard centuries later. This outside source is a profound mystery for which no rational explanation is either possible or necessary.

St. Augustine believed that the obsession of science—the belief that one can discover and understand truth by reason alone—has its roots in a passion for independence and self-sufficiency or in pride, which is the original sin. This means that the conditions of wisdom are fundamentally moral and not intellectual. Therefore, for the attainment of wisdom, the love of self must be subordinated to the love of God. However, such subordination is beyond the ability of the individual. What is required is divine grace. The doctrine of sin and grace most profoundly marks the breach between classicism and Christianity,[5] as it does between science and Christianity from St. Augustine's time forth.

Sounding quite modern, St. Augustine argued that science does not create. Rather it constructs, using the material of sense perception. Scientific reason comes into play when the mind directs its attention outward and addresses itself to organizing the material of the sensory world. Its working is therefore contingent on the assumption that this material will present itself in patterns the mind is capable of comprehending. But there is no way reason can verify this assumption. Further, the patterns that emerge are, in all cases, relative to the capacities of the observer. Because of this, the limitations of science are not merely those of the human faculties—the organs of sense perception and the information communicated from them, which one has to accept. The limitations are also of a creature immersed in the flux of time and space and swept along by a current whose velocity and direction the creature endeavors to chart. From these limitations there is no possibility of escape. In other words, there is no chance for anyone to exhibit so-called unbiased, objective, disinterested reason.

St. Augustine saw clearly that it is impossible for humans, by any effort of knowledge or imagination, to escape from the limitations of human nature and view things as they really are—in other words to discover and fully understand reality and truth. He believed that this realization should make us relinquish our aspirations to omniscience and recognize that our powers of apprehension are determined by the conditions of our existence as human beings, who are creatures of time and space. Essentially, he saw that we are all dancing in the dark, or to reference the words of St. Paul from the

Bible, of whom St. Augustine was a great scholar, we see reality and truth only as "through a glass, darkly" (1 Corinthians 13:12; King James Version).

So, a leap of faith is necessary if one is to apprehend truth and reality. For many scientists, both classical and modern, this leap is made, in large measure, to mathematical models and metaphysical abstractions. But essentially it is to faith in themselves and in their own reason as well as to the sovereignty and providence of reason alone. For St. Augustine and for the Church, it was to the sovereignty and providence of God, represented by the concept of the Trinity and demonstrated in the person of Jesus Christ. St. Augustine recognized that in making this leap of faith, he had passed beyond the point about which science can say anything. He understood that any attempt to describe the operations of Deity would involve the use of symbols, which are, of course, metaphors. It appears that the metaphors of quantum science have likewise passed beyond the point about which science can, or possibly should, say very much of anything.

Although St. Augustine and the Church believed that they had adequately addressed the problem of the One and the Many, quite obviously they had not—at least not for those outside of their faith and authority. The problem of the One and the Many remained an ongoing intellectual exercise and flowered again amongst philosophers in the West once the authority of the Church was undermined. Of course, the problem remains today. It became obvious again in the modern era when quantum mechanics, with its distinctly idealist bent, undercut the classical materialist theories of Newton. The search for a resolution continues as particle physicists try to come to grips with such concepts as string theory and whether elementary particles are, or are not "little things." This is a debate between the materialists and the idealists.

The differing perspectives concerning truth and reality and the resolutions of the problem of the One and the Many, as expressed by St. Augustine, the Christian Church, and classic Greek philosophy and science, has been described as a debate between faith and reason—Christian faith against Greek reason. I do not view it as such. In my opinion, a fuller perspective sees it as the use of reason to argue for two quite distinct, and ultimately metaphysical faith positions: faith in reason and faith in revelation.

Christianity won this debate with Greek science and, after the conversion of the Roman emperor Constantine, who recognized Christianity as a licensed cult with the Edict of Milan in AD 313 and the Council of Nicaea in AD 323, rose to become, in effect, the state religion. For over a thousand years it was the monopoly faith in the West. This was challenged during the Enlightenment with the rise of science and the resurgence of faith in reason alone. But both science and Christianity relied on reason and rational arguments to understand and elucidate their perspectives. Both repudiated the mystical cults that relied on feeling and apprehension to understand reality and truth.

So the contention of this chapter is that the debate between Christianity and Greek science—just like the present debates between religion and science—was, and is, a rational debate based on different faiths. In effect, although we may be dancing in the dark, we are all dancing in faith of one kind or another.

The one faith, that of the Greeks, natural philosophy, and science to this day, is that through our reason, and reason alone, we are able to discover what reality and truth are and that we will discover the One and its relation to the Many. This has been described and extolled as a disinterested rational approach, because it is said to not be based in any ideological or theological position.

The other, that of Christianity, is that—while reason must be utilized in the pursuit of knowledge and understanding—ultimate truth can never be discovered through reason alone, but must come as a gift from God, as revealed truth, since to know truth or the One is to know God. Reason is used to understand, to elucidate, and to prove revelation. This could be described as an interested rational approach, because it is based in a theological and/or ideological perspective.

Chapter Fourteen
The Victory of Disinterested Reason

Four great movements mark the decline of the Christian line of thinking that truth must be revealed and thus the Middle Ages. These movements led to the Enlightenment of the seventeenth century. They were all predicated upon the gradual rise of vernacular languages via rich oral traditions that competed with the religious truths and perspectives enshrined in Latin. The invention of the printing press in the middle of the fifteenth century had a truly revolutionary effect as it served to "broadcast" the vernacular languages and with them, different and competing theories and perspectives to masses of people.

The emphasis placed on Latin by the Church, which had allowed the Church to preserve and protect its religious truths, had created a wide—and growing—gap between religion and the various oral and later written vernacular traditions. The royal courts were important centers of literary activity in the vernacular and supported the poetry of the troubadours as well as the literature of writers such as Dante, Petrarch, Boccaccio, and Chaucer.

The influx of Greek scholars from the East, following the destruction of Constantinople in the middle of the fifteenth century, contributed to a growing interest in classical Greek culture and philosophy. The vitality of Greek classics gradually weakened the monopoly of truths held by the Church.

The first great movement was the Italian Renaissance of the fifteenth and sixteenth centuries. While Dante still reflected medieval ways of thinking, he wrote in the vernacular so the masses of people who did not know Latin could read his works. More secular ideas can be seen in the works of writers such as Boccaccio and Petrarch. As a result of these vernacular writings and others, there was a rebirth of interest in the secular, scientific outlook reflected in ancient Greek philosophy. Indeed, the Renaissance movement has been characterized as a return to the ancients.

Whereas medieval society and culture were dominated by religion and God, the Renaissance thinkers were more interested in Man. It is because of this perspective that the new cultural attitude was given the name of humanism, which was the second of the great movements. It differed from the Renaissance, which had a profound influence on people at large, in that it had a more tightly focused influence on thinkers and scholars. Within a century, the influence of the Renaissance and of humanism had moved north into Germany and to France and then to England.

In England, the humanist movement was contemporary with the Reformation, the third of the great movements—by far the most dramatic—that changed the medieval world. This movement brought forth both a religious and a political revolution, causing war and carnage throughout Europe. It begat Protestantism; destroyed the monopoly of knowledge of the Roman Catholic Church; established the notion that the individual could ascertain the truths—initially of the scriptures, but consequently of a great many other things—greatly influenced the spread of printing in the vernacular; and supported the movements associated with the evolution of national monarchies, secularism, and modernism.

The fourth movement was the revival of natural philosophy, powered by the resuscitation of Greek philosophy. This great movement, which became modern science, was indebted to the mathematical theories of Pythagoras and Plato, which led to the extraordinary developments of science in the sixteenth and seventeenth centuries with the works of Copernicus, Kepler, Galileo, and Newton and continue to this day.

The comparatively esoteric understanding of mathematics in this fourth movement gave man new power over reality and made him more like God. The Pythagoreans, of course, viewed God as the supreme mathematician, so insofar as man could improve and exercise his mathematical skills, the closer he came to divine status. Plato's Ideas and Forms were seen as mathematical structures, and it is these structures, these mathematical realities, that have become the pursuit of natural philosophy, of science, since that time.

As referenced, the religious protest movements of the fifteenth and sixteenth centuries, powered by the new technology of printing and the power of vernacular languages, undermined the power and authority of Rome and the Papacy. Protestantism introduced multiple authorities. With the principle of the "priesthood of all believers," authority was based on literal readings of the Bible. However, for practical and political purposes, power gravitated to organizations of believers, or churches. But there were many of those. Within virtually all Protestant organizations, philosophical speculation was tightly contained. Even so, overall singular control was destabilized, and the Protestant Reformation opened up the field for speculation that would have been impossible before it and, I would argue, tilled the ground for much broader speculation that developed with and around it.

Certainly in England and Holland in the wake of the Reformation, there developed a new attitude toward politics and philosophy. This became known as liberalism. While it was essentially Protestant, it was not narrow in its outlook, as much of Protestantism was. Rather it appears to have been a development of the belief that each man must believe in his own way. Since liberalism was also a product of the rising middle class, it was opposed to the conservative traditions of privilege of both the aristocracy and the monarchy. Additionally, perhaps because bigotry is bad for business, this movement promoted toleration.

In the seventeenth century, when most of Europe was

being torn apart by religious strife, the Dutch Republic was a refuge for nonconformists. Since the Protestant churches never acquired the solidarity and political power of Roman Catholicism, the secular power of the state became more important. The liberalism of the seventeenth century was therefore a force for genuine liberation. It was equally opposed to the dying traditions of medievalism and to the extremist Protestant positions. Protestant theology had emphasized, at least in principle, that matters of belief and of conscience were individual and not subject to any external authority.

René Descartes, writing in the seventeenth century, is regarded as the founder of modern philosophy. He certainly is a philosophical father of Western science. His method reflects the influence mathematics had on him. After analyzing how mathematics could lead to far-reaching consequences in other fields and thus enable one to reach similar kinds of certainties, which were not possible with any other means of inquiry, Descartes decided that he must look inward to find the truth. In a perspective that foreshadowed the developments of science, he rejected everything he had been forced to take on trust. Only logic, geometry, and algebra were left. While this worked in mathematics, it led to systematic doubt in metaphysics. He realized, as did philosophers before him and as do their heirs now, that the evidence of the senses is uncertain and must be questioned. Sounding very modern indeed, he believed that even mathematics must be suspect, since God might be systematically leading us astray.

Ultimately, he decided that the one thing that even a doubter must admit is his own doubting. This is the basis for the fundamental Cartesian formula, "I think, therefore I am." Here, he believed, was a clear and distinct starting point for metaphysics. The Cartesian philosophy, like Greek philosophy before it, emphasizes thoughts as the starting point. This, and his method of critical doubt, has influenced Western philosophy and science ever since.[1] In addition, it sharpened the old dualism between mind and matter—a dualism that was emphasized by science until the advent of quantum theory. The rigid determinism of the material world, especially when taken up in Newtonian physics, became a fundamental truth of science.

The works of Descartes influenced two major philosophic streams of development, the so-called rationalist tradition and British empiricism. The three great representatives of the latter were John Locke, George Berkeley, and David Hume, who all lived and worked in the seventeenth and eighteenth centuries. It is the works of Berkeley that I want to emphasize. However, the questions concerning reality and truth, which all three debated during this period, are very similar to the ones that were and still are debated throughout the history of the West.

Locke's most famous philosophic work is his *Essay Concerning Human Understanding*. In it, he attempts to elucidate the limitations of the mind and the sort of inquiries that are possible for us to pursue. Whereas the rationalists of the Greek tradition assumed that perfect knowledge was attainable, Locke was less optimistic. His thesis claimed that knowledge must be based strictly on experience. The mind begins as a clean sheet of paper. Experience provides it with mental content or ideas. This meant that the innate ideas of Descartes, as well as the Ideas of Plato, had to be rejected. Reflecting a similar thesis, some scientists still believe that the only information that can legitimately be called knowledge is that which can be validated through repeatable laboratory experimentation—in other words, by experience.

George Berkeley disagreed. His theory, classic idealism, stated that for something to exist, it is the same as that thing being perceived. He challenges his reader to find any other meaning for "the table exists" or "the horse is in the stable" than "the table or the horse is perceived or perceivable." To speak of anything as existing makes sense only in and through experiences. Therefore, to be and to be perceived are one and the same.

Yet Berkeley also asserted that his theory did not affect the reality of things, meaning it did not take away independent existence. He said that natural things do not depend on one's mind like the imaginary image of a unicorn does. However, being ideas, they cannot subsist by themselves. Therefore, there must be some other mind in which they exist. For him, this was the mind of God. In this way, the table is there and the horse is there even if nobody is there to perceive them because they are perceived by God. Berkeley's objective was to show that sensible things have no absolute existence independent of mind, therefore cutting the ground from under the feet of the materialists.[2]

As I have suggested earlier, while it does not propose God as the ultimate consciousness that perceives and thus creates reality, the Copenhagen school of quantum theory is closely aligned with Berkeley's philosophy and is the current exemplar of the philosophy of idealism. The Copenhagen interpretation of reality is that given by its founder, Niels Bohr, who believed that our sensory world of experience floats on a subatomic world that is not real; there is no quantum reality in the absence of observation, or observation creates reality.

Since our world is made up of elementary particles, which are like ghosts, the things of sensory perception cannot exist apart from being perceived. So the question arose: "Do you really believe that the moon is not there unless we are looking at it?" It became a serious issue for a debate that continues. What was so controversial about it when Berkeley proposed it, and remained so when quantum theory brought it again to the forefront of inquiry, is that the idea of perception as existence is so dramatically at odds with common sense.

Berkeley mounted a serious critique of some of the metaphysical assumptions of Newton's scientific theories, just as quantum mechanics mounted the serious critique and ultimately undermined the classical physics theories of Newton. But Berkeley did this for different reasons, peculiar for the age in which he lived. Basically, he criticized the growing materialism of the age, the belief that all reality was matter, that all metaphysical and religious speculation was nonsense, and that reason and science would usher in an earthly paradise. Essentially, this was the deification of reason and man.

Berkeley clearly saw that scientific theories are hypotheses, and he understood and tried to emphasize that it is a mistake to think that, because a scientific hypothesis works, it must, of necessity, be the expression of the human mind's ability to penetrate the ultimate nature of reality or truth. In remarkably prescient ways, his theories challenged the materialism of the classical physics of Newton from the very beginning. But few paid attention. Nor would they until the triumph of quantum theory.

David Hume was also skeptical of the scientific truths of his day, and it is because of his honest skepticism that he is referenced here. As I emphasized earlier, skepticism does not mean cynicism or some chronic inability to make decisions or to take a position. The original Greek simply meant "one who inquires with care."[3] The Greek skeptics questioned the truths of the metaphysical system-builders. Where the system-builders felt they had found their answers, the skeptics kept looking. Over time, it was their lack of confidence rather than their continuing questioning that did the Greek skeptics in and resulted in what has become a rather negative reputation. Hume's philosophy was skeptical in the original sense and left no reason for a lack of confidence or an inability to get on with ordinary pursuits.

It is this original sense of skepticism that is now being

reawakened by leaders in the scientific community. Hume, and modern scientists, came to the conclusion that certain things that we take for granted in everyday life cannot be justified, even those that are based on scientific truths. The aim of science is to exhibit causal relations within a deductive system where effects follow from causes as the conclusions to valid arguments follow from their premises. Hume's criticism, and those of the true skeptic, remains valid for the premises themselves.[4]

It is this—the questioning of basic premises—that caused the quantum revolution. One can argue that the questioning of basic premises causes every revolution, at least every intellectual and spiritual revolution. While the quantum revolution has been rather tightly contained within the circles of theoretical physics, because of the extremely esoteric arguments of mathematics, it is not impossible to imagine that a wider dissemination of quantum theories, using plain language, might promote a wider revolution of thought.

The German philosopher Immanuel Kant, in a huge body of work written in the eighteenth century, struggled with these same questions of how one can ascertain what is reality and truth. While he agreed that knowledge arises through experience, he added an important perspective: that we must distinguish between what actually produces knowledge and the form that that knowledge takes. While sense perception is necessary, it is not sufficient for knowledge. He believed that the principles of organization in the human mind that transform the raw material of experience into knowledge are not themselves derived from experience. This knowledge that is independent of experience he called *a priori*. That which is derived from experience he called *a posteriori* knowledge.

The objective of one of his major works, *The Critique of Pure Reason*, is to explain how *a priori* knowledge is possible. This was important to him because, among other things, he believed that purely mathematical propositions can be classified as *a priori* knowledge. In this, he reflected the beliefs of Plato and the Pythagoreans and passed on a perspective that was central to Newtonian physics and remains central even after the quantum revolution.

Kant was a firm believer in Newtonian science. The validity of this scientific conception of the world remained a firm fact. The nature of scientific knowledge was certainly open to question, and in the course of time, he found himself raising basic questions about the nature and validity of scientific knowledge. He certainly would have disagreed strongly with the generalization of scientific truths that tended to overpower critical thinking in succeeding years. But he never doubted the general validity of Newton's physics within its own field. His works served to philosophically justify Newtonian physics, which postulated the uniformity of nature. However, Kant understood that experience alone cannot prove this uniformity. It cannot show that the future will resemble the past in the sense of showing that there are universal and necessary laws of nature.

Therefore, the proof must be *a priori*; in this case, it was mathematics.

His arguments are complex and perhaps dated. But questions surrounding the basic presuppositions of natural philosophy and science since the time of Newton and during the time of quantum mechanics, along with what their logical status might be, are very current. Arguments still rage as to whether mathematical truths are in the nature of things to be discovered or if they are abstractions of the mind. Kant's critique of idealism, while specific to his historical context refuting the theories of Descartes and Berkeley, is also applicable to quantum theories.

Kant argued that internal experience is possible only through external experience. Within the sphere of empirical reality, we cannot give a privileged status to the empirical self as Descartes did, reducing external objects to representations of the empirical self. The empirical reality of the subject is inseparable from the empirical reality of the world. Self-consciousness is not *a priori* datum. A human being becomes conscious of himself or herself in perceiving external things.[5] So just like those theoretical physicists who criticize the Copenhagen school of quantum physics, Kant insisted on the empirical reality of the world of experience.

In addition, while he saw that metaphysics is possible as a natural disposition because of the very nature of human reason and that it was valuable as a natural tendency, as science, metaphysics was impossible. In a critique that remains very relevant, he said that metaphysics that employs the mathematical method, as does quantum theory, will be confined to exhibiting relations of formal implication (i.e., of mathematical forms). Kant believed that if the metaphysician really wishes to increase our knowledge of reality, he must cease trying to rely on mathematics and turn to the method used so successfully by Newton in natural science. The metaphysician should begin by clarifying the confused concepts of experience and give them adequate expression. Kant was not possessed of any great faith in metaphysics to extend our theoretical knowledge of reality beyond the sphere of the sciences.

Most relevantly for our discussion of quantum theory, his skepticism is linked to the conviction that, whereas natural science has made good its claim to increase our knowledge of the world, metaphysics has not yet done so. He speaks of it as "a bottomless abyss" and as "a dark ocean without shore and without lighthouses."[6] Similar criticism is increasingly being heard regarding areas of quantum mechanics such as string theory, because they are far beyond any kind of experimental verification.

If Immanuel Kant was the preeminent philosophical thinker of the eighteenth century, Georg Wilhelm Friedrich Hegel was that for the nineteenth. In him, German idealism and idealism generally received its ultimate form. German idealism survived and had great practical influence in the dialectic materialism of Karl Marx and Friedrich Engels and, in an even blacker practical manifestation, in the national socialism of Nazi Germany. As indicated earlier, idealism can and often does lead to monism—the belief that all thought or spirit is one—or pantheism—the belief that everything is spirit and is one. A clear manifestation of monism is prominent in the philosophy of Hegel.

Like all the great philosophers before him, he grappled with questions of finite and infinite, the changing and the unchanging, the Many and the One, or, in the words of the current similar debate in quantum theory, between sensory experience and unchanging reality or truth. He understood that if we put ourselves in the position of spectators, as science does, life appears as an infinite organized multiplicity of finite things (i.e., as Nature). Indeed, nature can be described as life posited for reflection or understanding. However, the individual things, organized in nature as a whole, are transitory and perishing. Thought, which is a form of life, is the unity between things and is an infinite, creative life that is free from the mortality that affects finite individuals. This creative life, which is real and not a mere abstraction, is called spirit or God. Yet, with this conceptual framework, the problem remains the same: the relation of the finite to the infinite, with the idea of the infinite as spirit.[7]

For Hegel, the fundamental purpose of philosophy is that of overcoming oppositions and divisions. Like theoretical physicists, he believed that the fundamental interest of reason is to attain a unified synthesis: the Absolute, in Hegel's terminology, and a theory of everything in modern terminology. The subject matter of Hegel's philosophy is the Absolute, and the Absolute is reality as a whole, or the whole universe. This whole is infinite life or the process of self-development. Human beings participate somehow in this spirit and are a part of it.[8] The objective of philosophy, for Hegel, is to comprehend this whole but to do it through reason and not through a mystical state. This is precisely what the theoretical physicists are trying to do. Both Hegel and theoretical physicists appear to see "all" as essentially spirit and see reason as the way to reach and comprehend reality and truth.

Bertrand Russell says that Hegelian philosophy is based on a principle that recurs throughout the history of philosophy that no portion of the world can be understood except in its setting in the universe as a whole. Therefore the whole is the only reality. This perspective is found even among the pre-Socratic philosophers. Essentially, it is the Many and the One debate. The theory of Parmenides that the universe is an immovable sphere expresses something of this kind. The mathematical philosophers of the Pythagorean School also suggest this notion when they say that all things are numbers. Finally, the mathematically focused theoretical physicists, in their search for the supreme formula that will explain the whole of the universe, are moved by the same belief.[9]

Soren Kierkegaard reacted against theories of everything and the grand generalizations of philosophy in general, but Hegel specifically. In doing so, he reacted against the faith his philosophical predecessors in the West held in the ability of man's reason to uncover reality and truth and reflected not only his own Christian faith but the more general critique that Christianity had always leveled against an exclusive faith in reason. Kierkegaard was also reacting against what he saw as the logical extension of the philosophy of idealism to a monistic or pantheistic view of reality that all is spirit or mind or God and that we are a part of and participate in this whole or universal spirit.

Kierkegaard believed that philosophy, with its emphasis on the universal rather than the particular, had tried to demonstrate that man realizes his true essence in proportion to his rise above mere particularity, becoming enmeshed, through his reason, in what is universal. This theory, he argued, is false, whether the universal is considered to be the state or the economic or social class or humanity or absolute thought or spirit.[10]

Kierkegaard was adamant that God is not man and man is not God and that the gulf between them cannot be bridged by rational dialectical thinking. That gulf can only be bridged by what he called "a leap of faith." This leap is a voluntary act by which man relates himself to God freely and individually as creature to Creator, as a finite individual to the transcendent Absolute. By this emphasis on the individual, on choice, and on self-commitment, Kierkegaard was making a practical attempt to get mankind to see the existential situation and the alternatives with which we are faced. It was the very opposite of other philosophic attempts to master all reality by thought and exhibit reality and truth as a necessary system of concepts. In this sense, it was a critique of classical Western philosophy and of much of modern quantum theory. The existence of God, of truth, has to be grasped existentially. No amount of demonstration can establish it. Kierkegaard, therefore, separates faith from reason, or at least faith from faith in reason.

Kierkegaard is associated with the birth of the existential movement in philosophy, which is a movement that emphasizes man's need to act and choose not because of philosophic reflection but from a spontaneous impulse of the will. Existentialism thus allows for faith, in the case of Kierkegaard, but does not predicate it. The principle of existentialism is sometimes stated as "experience is prior to essence": first, we know that something is, and afterward, what it is. This puts the specific before the general, and Aristotle before Plato. Kierkegaard puts will before reason and argues that we should not be too scientific, since science deals with what is general and only touches things from the outside.[11]

In light of the monumental intellectual achievements of Western philosophy, natural and otherwise, while Kierkegaard's specific critique, like that of St. Augustine, is important—even essentially important—in order to understand the nature of faith and reason and their differing perspectives and potentialities, it is my view that separating them is simply unreasonable. Faith—whatever that faith may be—always underlies reason, and reason always provides the pathway for faith—whatever that pathway may be. This is as true for science as it is for religion.

Chapter Fifteen

Conclusion: The Limits of Reason and of Certainty

As we have seen, the search for reality is a true waltz in wonder, for it is also the search for truth, for God, whatever metaphor we use to understand and describe our conception of what we believe to be ultimate and lasting. This is a historical search that has provided the motivation for philosophy, theology, mysticism, and science. In this search, theoretical physicists working their way through enormously complex and strange theories of quantum mechanics are engaged in what has impressed me as being the most interesting and potentially significant waltz of any of the groups traditionally interested in this sort of metaphysical speculation.

Quantum theory suggests a potentially revolutionary metaphysic, philosophy, and theology. It suggests that the basic elements of our world, those which make us what we are, are radically uncertain and that everything, or much of everything, we take for granted as real in our real world of sensory perception is illusion.

Whether or not the moon is actually there if nobody was around to see it, or indeed whether or not there is anything—any thing—if no human beings were there to perceive, is most certainly still a debatable question, but it is at least debatable. It was not debatable, at least not in respectable intellectual circles, when the theories of classical physics ruled our minds. There is now at least a respectable theoretical possibility that nothing would exist—no world, no universe, nothing—if we were not here to observe them. Perhaps, after all, humans are at the center of the universe—perhaps *are* the center of the universe.

On the basis of this theorizing, the answer to the fundamental question of "why is there something rather than nothing" is that there are human beings and human consciousness. This, of course, simply moves the question to "why are there humans and consciousness." But at least this puts human consciousness instead of merely "something" at the center of speculation and theory.

Certainly, quantum theory has mounted a serious critique of Western science since the time of Newton in its neglect of the human consciousness. It has shown that a person cannot be separated from nature, that the experimenter cannot be separated from the experiment, and that the so-called third-person objectivity of science is a myth. In this sense at least, quantum theory appears to be closer to the perspective of St. Augustine in that human consciousness, *logos,* or word, is at the center of any and all concepts of reality and truth.

While this most current manifestation of scientific inquiry is being pursued rationally and accurately within the traditional Western scientific faith that an ultimate reality or truth actually exists and can be apprehended by human reason, the reality that is being pursued may not be rational at all. It is uncertain; in the words of Einstein, it is "spooky" and thus most perplexing to scientists and the rest of us, who are the beneficiaries of a history of Western rationalism. Mystical connections have been suggested, but where this may lead is anybody's guess.

One direction might lead toward an understanding and acknowledgement of the limits of human reason. If common sense is radically misleading in assessing reality and truth, as relativity and quantum mechanics show that it is, why is reason not also misleading? If reality is deemed to be uncertain, even irrational, this has either to do with reality itself or the limitations of human reason. Any challenge to the fundamental faith of science in the preeminence of reason in the universe and the ability of human reason to uncover or discover reason would represent a revolution or reformation in our Western secular religion. But interestingly and perhaps significantly, such a reformation would also bring it closer to the traditional view of Western Christianity that, while reason must be used in the pursuit of the knowledge of truth and reality (i.e., of God), this knowledge is ultimately beyond human reason.

Scientists, like philosophers and theologians, have a very distinct and powerful faith perspective that is both a metaphysical and a theological perspective. They believe in an inherent order of things, that there is a reality or truth that underlies and provides the basis for our world of experience, and that there are natural laws that arise from this reality and that govern our world. They believe that these are worth discovering, and they believe in the ability of reason, and reason alone, to apprehend and understand them. In other words, scientists pursue truth in the faith that it exists and that it is worth pursuing. To again quote Albert Einstein: "There is no doubt that all but the crudest scientific work is based on a firm belief, akin to a religious feeling, in the rationality and comprehensibility of the world." This religious feeling is being shaken in the minds of at least some theoretical physicists by the apparent fundamental uncertainty of the quantum world.

Our secular society is not used to and, I suspect, will be very uncomfortable with uncertainty in science, just as the former religious society was uncomfortable with it in religion. However, uncertainty is not incompatible with a strong faith, as great thinkers in both religion and science have known. Faith can provide assurance without certainty. Additionally, being uncomfortable, at least intellectually, can be a boon to creativity. So long as we are able to withstand, and indeed revel in, uncertainty, quantum theories could motivate a truly protestant reformation (i.e., a protest against the old truths, which, like all truths, can and often do enslave the mind). While it will lead, in fact already has led, to some wild speculation given its spooky nature, this uncertainty may also lead to profound insights and perhaps some intellectual and emotional compromises and accommodations.

While the new science may undermine some traditional beliefs, both secular and sacred, it does leave wide open and encourages the faith that an ultimate reality and truth exists as the fundamental basis for the sensory manifestations of our physical world. Contemporary theoretical physicists, who are at the forefront of this new thinking, clearly believe this, as did their scientific and philosophical forebears in the West. Here they are also at one with their theological contemporaries and forebears. All have believed and now believe in the reasoned pursuit of intellectual understanding. They diverge, apparently, only in how this understanding ultimately comes about—the one believes it's through reason alone, the other via revelation or divine inspiration. But even here there is only a very fine distinction between the faiths. Science is very familiar with inspiration. Science just isn't prepared or able to award it divine status.

Great spiritual breakthroughs—great truths—are a combination of inspiration and perspiration. The great truths of science, after much discipline and hard work, often seem to come as a gift, an inspiration, or even a revelation. But this is not magic. Breakthroughs in physics and mathematics do not come to those who have not mastered the technique, just as great creative breakthroughs in art or music do not come to those who are neither accomplished artists nor musicians. The same is true for religion. All of these come from the faith that the truth—or the great art or performance or God—is "there" and it is possible to achieve or apprehend that truth. Without this faith, the hard work would not be worth it. Even so, importantly, hard work alone is not enough. True inspiration comes from inside or outside human consciousness, just as revelation does. The understanding or definition of where the inspiration comes from is a matter of faith.

So for science, as it always has been for Christianity, a leap of faith is necessary. For many scientists, classical and contemporary, this leap is made to mathematical models and metaphysical abstractions. But, again to emphasize, essentially the leap is to faith in themselves and their own reason and to the sovereignty and providence of reason alone. For the great mystics, it was to the emotional apprehension of an undifferentiated unity. The mystics understood that their apprehension was beyond all reason. For St. Augustine and the Church, the leap was to the sovereignty and providence of God. St. Augustine recognized in doing this that he and the Church had passed beyond the point about which science and human reason can say anything. He understood that any attempt to describe the operations of deity involves the use of symbols, which are, of course, metaphors. It appears that the symbols and metaphors of quantum science may have passed beyond that same point: beyond the limits of reason. So modern physicists, like their religious antecedents, may be left with faith alone. If this is the case, then secular society will also be left with faith alone.

The concept of salvation by faith alone is the classical basis for the Protestant Reformation. While physicists do not and would not use the word salvation to describe their goal of apprehending reality and truth, it is quite appropriate, in my opinion, for their goal is indeed a secular salvation. That is why the revolutionary quantum theories can be likened to the first Protestant Reformation.

The first Reformation, beginning in the sixteenth century, came about as a result of the translation of the Bible into vernacular languages and the mass dissemination of these Bibles through use of the printing press. The truths of scripture, up to that time, had been tightly controlled via the elite who could read Latin or Greek—or, heaven forbid, Hebrew or Aramaic. The scriptures were controlled by a priestly caste, just as the truths were written in hieroglyphic or hieratic characters and likewise controlled by priestly castes in ancient tradition.

Revolutions occur when new truths are communicated to the masses via commonly understandable metaphors. While the truths often, perhaps always, get changed, modified, or added to when this happens, the process is often liberating and sometimes destructive. In the case of the Reformation, once many could read and understand the truths of the Bible, there developed almost as many interpretations of these truths as there were readers. "The prolific source of Protestant sectarianism was the notion that the scriptures speak unmistakably."[1] But the freedom and democracy and modern society that arose from this revolution have been of inestimable value to mankind.

The same or similar benefits and dangers may come from the gradual translations of the esoteric mathematical metaphors of the new physics into the vernacular languages open to all readers. Such translations will probably lead to wild and wooly speculations and interpretations. But they may also lead, as did the Reformation, to profound insights and accommodations. The Protestant Reformation led to questions upon questions upon questions. Admittedly, many of the Protestant questioners, especially the early ones, were burned at the stake, but at least in the long term, the benefits far outweighed the costs—unless, of course, you were one of the ones who got burned. Although there are many current ways to get burned metaphorically, including being ridiculed, denied university tenure, and being fired, it is unlikely that such physical unintended consequences will attend this current intellectual and spiritual reformation.

This is because the process of questioning and undermining sacred truths has become well established within the science community since the time of the ancient Greeks. However, even within this general culture of questioning, just as the theological absolutists of the Church were appalled by translations of the scriptures into vernacular languages, many physicists will be appalled by opening up their truths to the mathematically and scientifically illiterate and the apostates. Many will ignore any outside writing and thinking as much as possible, maintaining their positions of intellectual and moral superiority over the "great unwashed." But, unlike the first reformation, there will likely be no violence and certainly no physical violence. There will be scientific sniping and defense mechanisms based on the esotericism of mathematics, but eventually accommodations will be made for other voices, and perhaps some scientific insights may even be gained from outside perspectives.

Just because something is spooky or appears to be beyond reason or has been given via revelation does not mean that its understanding should not be pursued rationally. In our brief look at St. Augustine and the early Church, we saw that, even though their truth was beyond reason, the understanding of it was pursued rationally. However, when revolutions and reformations occur, their results are unpredictable.

If, as the spark of the early Reformation once stated, "a man must do his own believing, as he does his own dying," then we should strive to have as many of the best and most durable beliefs about reality and truth as possible. Now, Luther would not have countenanced this. His Reformation consisted of truths against truths. This reformation could be very different, at least for a while, in that the *pursuit* of truth could be a real goal in the faith that it does exist and that it is worth pursuing, even though we may never know, verily know definitively (i.e., scientifically), what it is.

Just as the traditional religion of Christianity did before it, our secular religion of science has given us metaphors of certainty. The visual metaphor during the Middle Ages in the West was Dante's picture of reality as a series of concentric spheres—heaven at the top, next the planets' crystalline spheres, down through Earth's concentric elements, to the seven circles of hell. Newton's scientific revolution toppled all of this, and his clockwork universe was a mechanical model in which everything that happens has been predetermined by the laws of nature and how the universe was in the distant past. This model coincided with a practical, realist view of the world that reality can be perceived via the senses, is separate from our perception, and can be objectively studied. These classical truths have now been overturned. What quantum mechanics has shown us very clearly and precisely is that reality is far more complex and mysterious than many people, especially scientists, had imagined.

Still, what Newton provided, and what has influenced the West's view of the world for the last several hundred years, was a clear and certain metaphor of reality: the world as a giant clock that is perhaps wound up by the Creator but running independently. Modern quantum theory provides no such clear metaphor. Its metaphor—if one can call it that—appears to be uncertainty.

The definition of *metaphor* in *Webster's New World Dictionary* is "a figure of speech in which one thing is likened to another." A popular example is the Shakespearian claim that "all the world's a stage" (*As You Like It,* act 2, scene 7). The examples referenced above are reality as a mechanical clock or as a realm of uncertainty, a realm of pure energy that is sometimes even expressed as spirit. In effect, all the ways we express our human experience are metaphors—literally, a carrying over of signs, a stand-in for another reality. Language is metaphor, as is mathematics. We, nonscientists, communicate our understandings of reality, whether scientific or religious or mystical or philosophic, by using words. Our deepest realities force us into metaphor, which are often confusing, as words and metaphors about ultimate reality and truth are wont to be. This is why physicists prefer mathematics.

The high priests of our secular religion, the theoretical physicists, decry the nonmathematical speculations and metaphors of nonscientists about uncertainty and the uncertainty principle, which is the very heart of quantum theory. In general, they are just carrying on a priestly tradition in trying to keep esoteric knowledge esoteric. But in some cases I am sure they are justified. Perhaps in this case, but I hope not. At least one of their major concerns appears to be the misappropriation of the term uncertainty to justify inaccuracy, whereas it is a precise statement regarding measurable quantities.[2] Far be it from me, I hope, to do this. It is the very accuracy of the measurement of elementary particles in the quantum world that demonstrates the uncertainty of this world. And it is this uncertainty that has revolutionized science.

Like philosophers, theologians, even mystics, and the rest of us, scientists use metaphor to describe the reality they are searching for and/or dealing with. This is most obvious to us when they try to describe reality to nonscientists and nonmathematicians (i.e., when they are forced to use nonmathematical metaphors). By necessity, they must use mental pictures and physical analogies. However, many of them still disagree over whether such pictures are useful.

The uncertainty of such pictures is in much evidence when physicists try to describe the quantum world, which is a world that cannot be seen. Nobody has seen an electron. Scientists can see the tracks of an electron in an experimental cloud chamber or when an electron impinges onto a television-like screen. But nobody has actually seen one.

Niels Bohr took the position that "the quantum world does not actually exist." Other scientists have said that we can never know what an atom actually is, in reality. We can only know what an atom is like. Some physicists, Richard Feynman being one example, essentially gave up trying to describe any of this using commonsense analogies. Feynman is said to have remarked that he could picture invisible angels but not light waves.[3]

However, to speak at all about such things is to use metaphor. Cosmologists talk about the Big Bang as though this were some kind of massive explosion. Theoretical physicists of the new branch of quantum mechanics called string or superstring theory talk about these impossibly tiny "elements of energy" as actual strings, "stretching, vibrating, breaking." However, unlike the classic metaphor referenced above—"all the world's a stage"—which has a literal connection, metaphors used in the new theoretical physics have no literal element. When they reference a string vibrating, they don't even know if it is an actual string—a thing—let alone if it actually vibrates. If superstring theory requires ten or eleven dimensions, scientists, and we, have no idea at all what they might look like.

Even the comparatively simple descriptors are beyond us. We cannot visualize spacetime where time is the fourth dimension. When the universe is described as being shaped like the surface of a balloon, we know practically nothing about how space might curve in three dimensions, let alone in four. Of course, a massless particle is also difficult to visualize, although it is conceived as a definite piece of matter—a "little thing"—and described as being so small as to be considered as having no magnitude, only inertia and the force of attraction. Difficult indeed.

Clearly, all metaphors, certainly all metaphors using words, are inadequate in this impossibly complex, subatomic world. But then words have always been inadequate to describe ultimate reality and truth, however they may be envisaged.

As we have seen, currently the West has three secular metaphors for how the universe is conceived, plus the Judeo-Christian religious one: classical, or Newtonian, physics and Einstein's special and general relativity are both reasoned metaphors, and quantum mechanics, while certainly a reasoned metaphor, indicates a reality that is uncertain and beyond reason. The religious metaphor is also reasoned and points to a reality that is beyond reason. All of them, secular and religious, are attempts to apprehend, understand, and reconcile the One ultimate reality or truth that both science and religion believe exist with the Many realities or truths of our everyday existence.

While the classical science metaphor is outdated, we have seen how theoretical physicists have been working overtime to come up with a mathematical one: a wholly rational theory that will reconcile the relativity and quantum theories. This would provide one unified theory to elucidate ultimate reality in a mathematical equation. It would be a theory of everything and, of course, a final victory for the faith of science in reason, and reason alone, to uncover or discover and fully realize ultimate reality and truth—a final victory in the minds of mathematicians and theoretical physicists. We have seen that many physicists believe that some version of string or superstring or M-theory will mesh quantum theory and relativity into one unified theory that will describe the One reality and truth that is the basis of the Many.

However, at least to this point, the problems of the One and the Many continue. The first is a problem relating to the One itself, to the most fundamental, elementary particles or essences that are the basis of everything. The faith that there is a One or a reality beyond the Many impressions of the senses, has been maintained by the majority of voices in science, just as it has by religion. But in science, there has always been a debate about the nature of this One, whether it is form or matter. For religion, it has always been form. With the rise of science in the Enlightenment, matter became predominant. To quote again the nineteenth century German radical Ludwig von Feuerbach: "The old world (the world of Christianity) made spirit parent of matter. The new makes matter parent of spirit." Now, with quantum theory, it appears that spirit is again parent of matter, although there still seems to be some disagreement about this.

Another problem relates to the question of which came first: the One or the Many. This question goes back at least as far as the different perspectives taken by Plato and Aristotle. For Plato, the One is paramount and all manifestations of the Many emanate from it. For Aristotle, the Many are paramount and our intuitions and conceptions of the One come from our experiences of the Many. In general, religion has sided with Plato, and critics of religion with Aristotle. In religious terms, the question boils down to whether man is made in the image of God—from the One to the Many—or God is made in the image of man—from the Many to the One. In scientific terms, it is a question of whether or not there is a reality or a truth outside of and beyond any human conception of it. A basic faith of science is that there is. Laws of nature and mathematical truths are there to be discovered, as is the One final truth: reality itself. Quantum theory supports this metaphysical perspective.

Reality or the One for string theory appears to be energy, pure energy that evolves into matter or the Many. The fathers of quantum theory (for example, Niels Bohr and Werner Heisenberg) clearly did not believe even then, before the development of string theory, that elementary particles are little things. Bohr was quite specific about this. Both believed that the smallest units of matter are, in fact, not physical objects in the ordinary sense of the word; they are forms, structures, or—in Plato's sense—Ideas, which can be unambiguously spoken of only in the language of mathematics.

It is really the beauty of the math and of the equations that is the basis of quantum theory. Theoretical physicists are treading uncharted metaphysical waters, with only their confidence in the logic of mathematics and their faith in reason to support them.

As we have seen, the scientific worldview in the West (science's faith in mathematics and reason alone) has its roots in ancient Greece. St. Augustine dismissed the faith of classical Greek science, whether idealism or materialism, with the phrase, "only their ashes survive."[4] One is inclined to respond "some ash, some survival," since not only was this the faith of Greek science, including Plato, but it has been and remains the faith of Western science to the present day. The challenge for modern quantum theorists, which also faced Plato and Western Christianity, is to build a bridge of understanding between the Many and the One.

The method Plato proposed for doing this was that of a reasoned, verbal dialectic. Quantum theorists are proposing an equivalent but primarily mathematical dialectic, by which reason elevates itself from the illusory world of the senses to that of the higher forms or patterns or equations. These are the truly real, or certainly what may be called the archetypes of reality. From them one's mind is continually raised to the Absolute principle—theory, equation, Good, God; certainly the truth and ultimate reality that lies behind them. For both Platonism and modern theoretical physics, the problem is one of transcendence: both try to find a means to pass to some sort of heavenly place.

For the philosopher, according to Plato, the mind moves metaphorically upward from many specific things in the world to their conceptual Ideas or Forms, and still upward to the highest concept that is the Good—or the One or God. Similarly, for the modern theoretical physicist, there is the faith that the mind moves ever upward through disinterested reason and the knowledge of mathematical forms, which they believe are real and not merely figments of the imagination, to an apprehension and understanding of reality and truth, which is the One master mathematical formula.

The danger in this process is that it can easily slip into mysticism. It did just this in the final elaboration of Platonism by Plotinus. He shifted—or perhaps was forced to shift by the limitations of pure reason, certainly by the limitations of his reason—from the objective and rational to the intuitive and mystical aspects of Plato's philosophy. In so doing, it has been suggested, he underscored the central deficiency of classical science and thus of modern science: the limitations of human reason in attaining a vision of ultimate reality or truth and, coincidentally, of thus achieving what may be described as salvation, which is to be achieved through the intensive cultivation of the speculative faculty, through reason alone. Plotinus carried this to the point, perhaps the inevitable point, where it yielded an ecstatic vision of the One—a mystical One that lies beyond reason and beyond existence.

Will this same dead end become the end in quantum theory? Will it demonstrate the inability of reason alone to apprehend reality and truth? The evolution from the rationalism of Platonism to Neoplatonic mysticism involved the admission, for Plotinus and others, that there was an impossible gap, certainly an impossible rational gap, between the scientific and the superscientific intelligence. There was no way that reason alone could achieve a vision or understanding of reality or truth. For Plotinus, the One had ceased to be a possible object of thought and had evolved into an object of adoration. Could such a thing happen with the theory of everything—to the beautiful master equation or equations—so adored by theoretical physicists? Has it already happened?

Clearly, the history of Western Christianity indicates that this is in no way inevitable. From the very beginning, the fathers of the Church fought against this slippery slope. St. Augustine was a great scholar of both Plato and St. Paul. He and the other fathers refused to accept that the slide to mysticism was inevitable, even while they understood that human reason has its limits. To counter this slide, they formulated a rational theology based on a fundamental faith in revelation and backed by an overwhelming intellectual authority. A similar resolution may be found to counter any current slide: a fundamental faith in reason backed by an overwhelming intellectual authority.

Because of the weirdness and even irrational nature of the quantum world, there is a tendency, which is perhaps natural, on the part of some to try to understand it in mystical terms. This is understandable and may even offer some insights, but in the opinion of this writer, it is ultimately a dead end, at least for those of us in the Western tradition. Mysticism is essentially an Eastern approach to the discovery of truth, not a Western approach. To reach reality and truth, it goes beyond what is rational to what appears to be essentially emotional and irrational.

If not entirely irrational, mysticism does not use reason as its primary tool. Its entire premise requires that one go beyond reason to an emotional apprehension of what is real. Because of this, it says it is unable to explain or even express this apprehension rationally. But the entire history of philosophy, science, and religion in the West argues against this kind of irrationality. And reasonably so. After all, if truth is beyond reason, why is it not also beyond experience or emotion? The emotional mystical apprehension may also be *fantastica fornicatio*, or fornicating with one's own fancies. Of course it is possible that all apprehensions of what is ultimate truth and reality may be this. Or not.

While leading physicists have recognized from the very beginning the danger of the irrational nature of quantum reality, they continue to deal with it in the best tradition of Western rationalism. Scientists are fundamental, in fact profoundly fundamentalist, in their faith in reason. As for the rest of us, whatever our intellectual backgrounds and whether or not we share scientists' faith in human reason or in an ultimate reality or truth, we must not leave this exciting intellectual field to them alone. We must not back off simply because we do not understand their esoteric language, mathematics, or the intricacies of their theoretical constructs. We must press the physicists to translate their metaphors into ours, as some of them do very successfully, and then we must open our minds to understand them.

This is a real challenge, and a dangerous one for physicists and for science. It is much safer for them to maintain their intellectual isolation within the protective cover of complex mathematical formulae and "preach to the choir," as the saying goes. Even though there is plenty of criticism within their choir, so long as the math remains beautiful, there is comparatively little out-and-out derision, regardless of the weirdness of the theories presented. However, once scientists actually explain, in layman's words, what their theories are suggesting, there is a huge potential for the bullshit factor to gain prominence. The clearer they are able to be in their explanations, in some cases at least, the greater the potential for bullshit. In effect, they will face the same kinds of questions faced by metaphysicians, theologians, philosophers, et al., and unlike their scientific counterparts of the past, they no longer have the traditional fail-safe assurance of scientific truth and repeatable laboratory testing.

But, to explain themselves poses a risk not only for the scientists but also for the rest of us. Ideas, concepts, theories, and metaphors that might once have been described as science fiction are now under serious consideration at the leading edge of science, that of theoretical physics. New theories about strings and branes and extra dimensions of space and time have radically changed how physicists and cosmologists think about the world and reality. As these theories seep into the mind of the general public—if they do—changes in worldview may occur there also to an unknown effect. Principles of uncertainty are not the sort of thing to build public or individual confidence. While it may be a basic tenet of philosophy since the time of Socrates and Plato that we know nothing, it is not a basic tenet of human psychology or of politics. The public's faith in science is based on a belief that we know a lot. This is a vain faith, perhaps, but a strong one nonetheless.

In my opinion, they (scientists) and we must take the risk, for their very claims are among the most creative, exciting, illuminating, and magical of the modern world. They can provide unlimited avenues for intellectual experimentation, open wide and wonderful opportunities for philosophical, theological, metaphysical thought and speculation, and encourage thinkers of all backgrounds to creatively and intelligently consider our ideas of the nature of the universe, of man, of truth and reality, of God, in a fresh way and to search for new and more adequate images and metaphors.

In our very brief review of some Western philosophers, we saw that Soren Kierkegaard reacted against theories of everything and grand generalizations, specifically those of Hegel. But, in general, he was reacting against what he saw as the logical extension of reason in the philosophy of idealism to a monistic or pantheistic view of reality, which became an essentially mystical view—that all is spirit or mind or God or the One and that we, the Many, are a part of and participate in this One, or universal spirit.

However, underestimating reason is just as dangerous as overestimating it. Hegel tended to overestimate reason, Kierkegaard to underestimate it. Science tends to overestimate it, religion to underestimate it. In my view, all limit their potential by doing so. Quantum theory may provide an opportunity for science to gain a better balance. The extraordinary speculations of quantum physics undermine the old mechanistic truths of science and the old secular truths of the West and may force the secular mind to open once again to the central role of faith and to an understanding that faith and reason are not inimical but rather are complementary.

We may continue to have faith that reason can uncover or discover truth and reality, but we cannot know—in the old scientific and secular sense of knowing—that this is the case. Thus, as our traditional religion was undermined by classical physics our secular religion has been undermined by quantum mechanics. So our secular religion is now in relatively the same position as our traditional religion: a reliance on faith and reason rather than on experimental evidence.

Now for enlightened religious leaders—witness the words of St. Augustine and one would assume many, even most, theoretical physicists—this truth has always been self-evident. But for most of the adherents to the religion of science, as to the old religion, this has been anything but. Quantum theory, if translated well into words that are understandable to the many, could unsettle this complacency.

Whatever perspective we take, whether it is that reason alone is capable of discovering and understanding truth and reality, that only revelation can provide this fundamental understanding, that it can be reached only by a mystical apprehension, or that it can be only pursued and never reached or comprehended, or that there is no truth or reality to reach, it is a faith perspective.

Niels Bohr understood that all truths, science and otherwise, at least in part, are subjective in the sense that they all involve our particular perceptions and understandings. By definition, this makes so-called objective reality and life itself uncertain, given that our perceptions vary.

St. Augustine understood as well that it is impossible for humans, by any effort of knowledge or imagination, to escape from the limitations of their human nature or to view things objectively as they really are. He believed that this realization should make us relinquish our aspirations to omniscience and recognize that our powers of apprehension are determined by the conditions of our existence as human beings, as creatures of time and space. Essentially, he saw clearly, as did Bohr, that we are all dancing in the dark and we see reality and truth only "through a glass, darkly."

It appears that, perhaps for the first time in centuries or perhaps for the first time ever, the most profound metaphysical and theological speculations of science and religion in the West have, to some extent at least, coincided. For both, the world, reality, truth, God is mysterious and appears to be beyond human reason and human knowledge. To understand that we don't know, but to believe—as do both science and religion—that we can know at least in part, to believe that these ideas—reality, truth, God, whatever metaphors we use—are real and that the search for knowledge about them is worthwhile and valuable, is a waltz in wonder and a lesson to live by. It is also a lesson in humility. For it requires that we understand that we all live and die in faith of one kind or another, that we do not and cannot know what truth is, and that, in any event, it is not our truth but the pursuit of truth that makes us free.

Endnotes

Introduction

1. Arthur Schwartz/Howard Dietz, "Dancing in the Dark" (New York: Rhino/Wea, 1931).
2. John Barrow and Frank Tipler, *The Anthropic Cosmological Principle* (New York: Oxford University Press, 1986), 103.
3. Robert Laughlin, *A Different Universe* (Cambridge: Basic Books, 2005), 13–4.
4. Jacques Ellul, "Modern Myths," *Diogenes Magazine* 23 (1958): 25.
5. Julian Barbour, *The End of Time* (Oxford: Oxford University Press, 1999), 229.
6. Ambrose Bierce, *The Devil's Dictionary* (New York: Dover Publications, 1958), 107.

Chapter One

1. Brian Greene, *The Fabric of the Cosmos* (New York: Random House, 2004), 5.
2. K. C. Cole, *Mind over Matter* (Toronto: Mariner Books, 2003), 3.

3. Lisa Randall, *Warped Passages* (New York: HarperCollins, 2005), 115.
4. Michael Lockwood, *The Labyrinth of Time* (Oxford: Oxford University Press, 2005), 282.
5. Paul Davies, *About Time* (New York: Simon & Schuster, 1986), 163.
6. Dan Falk, *Universe on a T-Shirt* (Toronto: Arcade Publishing, 2002), 132.
7. K. C. Cole, *Mind over Matter* (Toronto: Mariner Books, 2003), 109.
8. Lisa Randall, *Warped Passages* (New York: HarperCollins, 2005), 2.
9. Ibid., 7
10. E. A. Burtt, The Metaphysical Foundations of Modern Science (New York: Anchor Books, 1954), 24.
11. Nick Herbert, *Beyond the New Physics* (New York: Anchor Books, 1987), xi.
12. Michael Lockwood, *The Labyrinth of Time* (Oxford: Oxford University Press, 2005), 294.
13. Shimon Malin, *Nature Loves to Hide* (Oxford: Oxford University Press, 2001), 145.
14. John Lukacs, *At the End of an Age* (New Haven: Yale University Press, 2002), 130.
15. Robert Laughlin, *A Different Universe* (Cambridge: Basic Books, 2005), 13–4.
16. Nick Herbert, *Beyond the New Physics* (New York: Anchor Books, 1987), 15.
17. Alan Lightman, *A Sense of the Mysterious* (New York: Pantheon Books, 2005), 42.

Chapter Two

1. Robert Laughlin, *A Different Universe* (Cambridge: Basic Books, 2005), 23–4.
2. E. A. Burtt, *The Metaphysical Foundations of Modern Science* (New York: Anchor Books, 1954), 31.
3. Ibid., 32–3.
4. Ibid., 226.
5. Brian Greene, *The Fabric of the Cosmos* (New York: Random House, 2004), 49.
6. Ibid., 67–8.
7. Lisa Randall, *Warped Passages* (New York: HarperCollins, 2005), 98–9.
8. Michael Lockwood, *The Labyrinth of Time* (Oxford: Oxford University Press, 2005), 79.
9. Karen Wright, *Discover Magazine*, Special Einstein Edition, September 2004, 50–53.
10. Michael Lockwood, *The Labyrinth of Time* (Oxford: Oxford University Press, 2005), 295–6.
11. Shimon Malin, *Nature Loves to Hide* (Oxford: Oxford University Press, 2001), 39.
12. Brian Greene, *The Fabric of the Cosmos* (New York: Random House, 2004), 80.
13. Ibid., 80.
14. Ibid., 139.
15. Lee Smolin, *The Trouble with Physics* (Boston: Houghton Mifflin, 2007), 6.

16. Julian Barbour, *The End of Time* (Oxford: Oxford University Press, 1999), 252.
17. Paul Davies, *God and the New Physics* (New York: Simon & Schuster, 1986), 107.
18. Lisa Randall, *Warped Passages* (New York: HarperCollins, 2005), 118–9.

Chapter Three

1. Nick Herbert, *Beyond the New Physics* (New York: Anchor Books, 1987), 20–26.
2. Shimon Malin, *Nature Loves to Hide* (Oxford: Oxford University Press, 2001), 111–3.
3. Ibid., 131–5.
4. Ibid., 227.
5. Ibid., 202.

Chapter Four

1. Brian Greene, *The Fabric of the Cosmos* (New York: Random House, 2004), 78.
2. Ibid., 17.
3. Ibid.
4. Robert Laughlin, *A Different Universe* (Cambridge: Basic Books, 2005), 7.
5. Brian Greene, *The Fabric of the Cosmos* (New York: Random House, 2004), 18.
6. Tom Siegfried, *Strange Matters* (New York: Berkley Books, 2002), 200.
7. Shimon Malin, *Nature Loves to Hide* (Oxford: Oxford University Press, 2001), 205.

8. Brian Greene, *The Fabric of the Cosmos* (New York: Random House, 2004), 354.
9. Ibid., 42.
10. Tom Siegfried, *Strange Matters* (New York: Berkley Books, 2002), 182.
11. Lisa Randall, *Warped Passages* (New York: HarperCollins, 2005), 293.
12. Ibid., 438–9.
13. Michael Lockwood, *The Labyrinth of Time* (Oxford: Oxford University Press, 2005), 92.
14. Paul Davies, *God and the New Physics* (New York: Simon & Schuster, 1986), 39.
15. Lisa Randall, *Warped Passages* (New York: HarperCollins, 2005), 454.

Chapter Five

1. Michael Lockwood, *The Labyrinth of Time* (Oxford: Oxford University Press, 2005), 114.
2. Tom Siegfried, *Strange Matters* (New York: Berkley Books, 2002), 107–8.
3. John Brockman, *The New Humanists* (New York: Sterling, 2003), 354.
4. Ibid., 356.
5. Ibid., 287.
6. Ibid., 299.
7. Tom Siegfried, *Strange Matters* (New York: Berkley Books, 2002), 259.
8. Robert Laughlin, *A Different Universe* (Cambridge: Basic Books, 2005), 14.

9. Lee Smolin, *The Trouble with Physics* (Boston: Houghton Mifflin, 2007), xiii.

Chapter Six

1. Ian Stewart, *Why Beauty Is Truth* (New York: Basic Books, 2007), x, 118.
2. Alan Lightman, *A Sense of the Mysterious* (New York: Pantheon Books, 2005), 65.
3. Paul Davies, *God and the New Physics* (New York: Simon & Schuster, 1986), 17.
4. Ibid.
5. Roger Penrose, *The Emperor's New Mind* (Oxford: Oxford University Press, 2001), 123–4.
6. K. C. Cole, *Mind over Matter* (Toronto: Mariner Books, 2003), 139.
7. Lisa Randall, *Warped Passages* (New York: HarperCollins, 2005), 104.
8. John Lukacs, *At the End of an Age* (New Haven: Yale University Press, 2002, 112.
9. Ian Stewart, *Why Beauty Is Truth* (New York: Basic Books, 2007), xiii, 276, 278.
10. Lee Smolin, *The Trouble with Physics* (Boston: Houghton Mifflin, 2007), 194–5.
11. Ian Stewart, *Why Beauty Is Truth* (New York: Basic Books, 2007), 223.
12. John Brockman, *The New Humanists* (New York: Sterling, 2003), 358.
13. Ibid.

14. Julian Barbour, *The End of Time* (Oxford: Oxford University Press, 1999), 23–4.
15. Dan Falk, *Universe on a T-Shirt* (Toronto: Arcade Publishing, 2002), 185.
16. Ibid., 186.
17. John Barrow & Frank Tipler, *The Anthropic Cosmological Principle* (Oxford: Oxford University Press, 1986), 123.

Chapter Seven

1. Evelyn Underhill, *Mysticism* (London: Methuen & Co. Ltd., 1960), 10.
2. Ibid., 3–4.
3. Ibid., 14.
4. Ibid., 22.
5. William James, *The Varieties of Religious Experience* (New York: The Modern Library, 1929), 371.
6. Evelyn Underhill, *Mysticism* (London: Methuen & Co. Ltd., 1960), 81.
7. Walter Stace, *The Teachings of the Mystics* (New York: The New American Library, 1960), 14–5.
8. Friedrich von Hugel, *The Mystical Element of Religion* (London: James Clark & Co. Ltd.,1961, 283–4.
9. Walter Stace, *The Teachings of the Mystics* (New York: The New American Library, 1960), 12.
10. F. C. Happold, *Mysticism* (Baltimore: Penguin Books, 1963), 18–9.

11. Walter Stace, *The Teachings of the Mystics* (New York: The New American Library, 1960), 23.
12. F. C. Happold, *Mysticism* (Baltimore: Penguin Books, 1963), 26.

Chapter Eight

1. F. C. Happold, *Mysticism* (Baltimore: Penguin Books, 1963), 28.
2. Shimon Malin, *Nature Loves to Hide* (Oxford: Oxford University Press, 2001), 115–6.
3. Ibid., 116.
4. Ibid., 119.
5. Ibid., 156–7.
6. Evelyn Underhill, *Mysticism* (London: Methuen & Co. Ltd., 1960), 23.
7. Ibid., 45.
8. Ibid., 24.

Chapter Nine

1. Hilda Graef, *The Story of Mysticism* (New York: Doubleday, 1965), 185.
2. Rufus Jones, *The Flowering of Mysticism* (New York: MacMillan Co., 1939), 9.
3. R. A. Vaughn, *Hours with the Mystics* (London: Strahan and Company Ltd., 1879), 278.
4. Ibid., 280.
5. Evelyn Underhill, *Mysticism* (London: Methuen & Co. Ltd., 1960), 96–101.
6. R. A. Vaughn, *Hours with the Mystics* (London: Strahan and Company Ltd., 1879), 210.

7. R. C. Zaehner, *Mysticism: Sacred and Profane* (Oxford: Oxford University Press, 1961), 205.
8. James Clark, *Meister Eckhart* (Toronto: Thomas Nelson and Sons Ltd., 1957), 23.
9. W. R. Inge, *Christian Mysticism* (New York: The World Publishing Co., 1964), 150.
10. John Noss, *Man's Religions*, 126.
11. R. C. Majumdar, *History and Culture of the Indian People* (London: George Allen and Unwin Ltd., 1951), 362.
12. Ibid., 467.
13. John Noss, *Man's Religions*, 126.
14. R. C. Majumdar, *History and Culture of the Indian People* (London: George Allen and Unwin Ltd., 1951), 493.
15. John Noss, *Man's Religions*, 126.
16. Ibid., 156–182.

Chapter Ten

1. Fritjof Capra, *The Tao of Physics* (London: Flamingo, 1991), 315.
2. Ibid., 224–5.
3. Ibid., 249.
4. Dalai Lama, *The Universe in a Single Atom* (New York: Morgan Read Books, 2005), 46–7.
5. Ibid., 56.
6. Ibid., 63–4.
7. Ibid., 66.
8. Fritjof Capra, *The Tao of Physics* (London: Flamingo, 1991), 209–10.

9. Dalai Lama, *The Universe in a Single Atom* (New York: Morgan Read Books, 2005), 92–3.
10. Ibid., 131.
11. Roger Penrose, *The Emperor's New Mind* (Oxford: Oxford University Press, 2001), 492.
12. Julian Barbour, *The End of Time* (Oxford: Oxford University Press, 1999), 26.

Chapter Eleven

1. William Braden, *The Private Sea* (Chicago: Quadrangle Books, 1967), 27.
2. T. Leary and R. Metzner, *The Psychedelic Reader* (New York: University Books, 1965), 191–2.
3. Walter Pahnke, *Drugs and Mysticism* (Cambridge: Harvard University, 1963), 46, 81.
4. T. Leary, R. Metzner, R. Alpert, *The Psychedelic Experience* (New York: University Books, 1966), 11.
5. William Braden, *The Private Sea* (Chicago: Quadrangle Books, 1967), 30.
6. Sidney Cohen, *The Beyond Within* (New York: Atheneum, 1966), 40.

Chapter Twelve

1. Harold Innis, *Empire and Communications* (Oxford: Clarendon Press, 1950), 80.
2. Ibid., 68.
3. Frederick Copleston, *A History of Philosophy*, Vol.1, Part 1 (New York: Image Books, 1965), 58.

4. Ibid., 80.
5. Roger Penrose, *The Emperor's New Mind* (Oxford: Oxford University Press, 2001), 146.
6. Frederick Copleston, *A History of Philosophy*, Vol.1, Part 1 (New York: Image Books, 1965), 203.
7. Bertrand Russell, *Wisdom of the West* (New York: Crescent Books, 1959), 82.
8. Frederick Copleston, *A History of Philosophy*, Vol.1, Part 2 (New York: Image Books, 1965), 31.
9. Bertrand Russell, *Wisdom of the West* (New York: Crescent Books, 1959), 98.
10. Ibid., 101.
11. Ian Stewart, *Why Beauty Is Truth* (New York: Basic Books, 2007), v.

Chapter Thirteen

1. Charles Norris Cochrane, *Christianity and Classical Culture* (Oxford: Oxford University Press, 1957), 422–3.
2. Ibid., 425.
3. Ibid., 410.
4. Ibid., 436.
5. Ibid., 451.

Chapter Fourteen

1. Bertrand Russell, *Wisdom of the West* (New York: Crescent Books, 1959), 195–6.
2. Frederick Copleston, *A History of Philosophy*, Vol.5, Part 2 (New York: Image Books, 1965), 49.
3. Bertrand Russell, *Wisdom of the West* (New York: Crescent Books, 1959), 229.
4. Ibid.
5. Frederick Copleston, *A History of Philosophy*, Vol.6, Part 2 (New York: Image Books, 1965), 67–8.
6. Ibid., Vol.6, Part 1 (New York: Image Books, 1965), 219–23.
7. Ibid., Vol.7, Part 1 (New York: Image Books, 1965), 201–2.
8. Ibid., 202.
9. Bertrand Russell, *Wisdom of the West* (New York: Crescent Books, 1959), 252–3.
10. Frederick Copleston, *A History of Philosophy*, Vol.7, Part 2 (New York: Image Books, 1965), 111.
11. Bertrand Russell, *Wisdom of the West* (New York: Crescent Books, 1959), 255.

Chapter Fifteen

1. Harold Innis, *Empire and Communications* (Oxford: Clarendon Press, 1950), 184.
2. Lisa Randall, *Warped Passages* (New York: HarperCollins, 2005), 117.
3. Alan Lightman, *A Sense of the Mysterious* (New York: Pantheon Books, 2005), 61.
4. Charles Norris Cochrane, *Christianity and Classical Culture* (Oxford: Oxford University Press, 1957), 165.

Bibliography

Augustine, St. *City of God.* Garden City, New York: Image Books, 1958.

Barbour, Julian. *The End of Time.* Oxford: Oxford University Press, 1999.

Barrow, John and Frank Tipler. *The Anthropic Cosmological Principle.* Oxford: Oxford University Press, 1986.

Bierce, Ambrose. *The Devil's Dictionary.* New York: Dover Publications, 1958.

Braden, William. *The Private Sea.* Chicago: Quadrangle Books, 1967.

Brockman, John, ed. *The New Humanists: Science at the Edge.* New York: Sterling, 2003.

Burtt, E. A. *The Metaphysical Foundations of Modern Science.* New York: Anchor Books, 1954.

Capra, Fritjof. *The Tao of Physics,* 3rd edition. London: Flamingo, 1991.

Clark, James M. *Meister Eckhart.* Toronto: Thomas Nelson and Sons Ltd., 1957.

Cochrane, Charles Norris. *Christianity and Classical Culture.* Oxford: Oxford University Press, 1957.

Cohen, Sidney. *The Beyond Within.* New York: Atheneum, 1966.

Cole, K. C. *Mind over Matter: Conversations with the Cosmos.* Toronto: Mariner Books, 2003.

Copleston, Frederick. *A History of Philosophy,* Volume 1: Parts 1 and 2, Volume 2: Part 1, Volume 5: Part 2, Volume 6: Parts 1 and 2, Volume 7: Parts 1 and 2. New York: Image Books, 1965.

Dalai Lama. *The Universe in a Single Atom.* New York: Morgan Read Books, 2005.

Davies, Paul. *God and the New Physics.* New York: Simon & Schuster, 1986.

———. *About Time: Einstein's Unfinished Revolution.* New York: Simon & Shuster, 1996.

Ellul, Jacques. "Modern Myths." *Diogenes Magazine* 23 (1958), 23–40.

Falk, Dan. *Universe on a T-Shirt: The Quest for the Theory of Everything.* Toronto: Arcade Publishing, 2002.

von Feurbach, Ludwig. *The Essence of Christianity.* New York: Harper & Row, 1957.

Feynman, Richard. *The Meaning of It All: Thoughts of a Citizen-Scientist.* Reading, Massachusetts: Perseus Books, 1998.

Graef, Hilda. *The Story of Mysticism*. New York: Doubleday, 1965.

Greene, Brian. *The Fabric of the Cosmos*. New York: Random House, 2004.

Happold, F. C. *Mysticism: A Study and an Anthology.* Baltimore: Penguin Books, 1963.

Herbert, Nick. *Quantum Reality: Beyond the New Physics, an Excursion into Metaphysics and the Meaning of Reality.* New York: Anchor Books, 1987.

Inge, W. R. *Christian Mysticism*. New York: The World Publishing Co., 1964.

Innis, Harold. *Empire and Communications*. Oxford: Clarendon Press, 1950.

James, William. *The Varieties of Religious Experience*. New York: The Modern Library, 1929.

Jones, Rufus M. *The Flowering of Mysticism: The Friends of God in the Fourteenth Century*. New York: MacMillan Co., 1939.

Laughlin, Robert. *A Different Universe*. Cambridge: Basic Books, 2005.

Leary, T., R. Metzner, and R. Alpert. *The Psychedelic Experience*. New York: University Books, 1966.

Leary, T., R. Metzner, and G. Weil, eds. *The Psychedelic Reader*. New York: University Books, 1965.

Lightman, Alan. *A Sense of the Mysterious: Science and the Human Spirit*. New York: Pantheon Books, 2005.

Lockwood, Michael. *The Labyrinth of Time.* Oxford: Oxford University Press, 2005.

Lukacs, John. *At the End of an Age.* New Haven: Yale University Press, 2002.

Majumdar, R. C., ed. *History and Culture of the Indian People,* Vol. 1. London: George Allen and Unwin Ltd., 1951.

Malin, Shimon. *Nature Loves to Hide.* New York: Oxford University Press, 2001.

Noss, John. *Man's Religions.* New York: The Macmillan Company, 1960.

Pahnke, Walter N. *Drugs and Mysticism.* (unpublished PhD thesis). Cambridge: Harvard University, 1963.

Penrose, Roger. *The Emperor's New Mind.* Oxford: Oxford University Press, 1998.

Randall, Lisa. *Warped Passages: Unraveling The Mysteries Of The Universe's Hidden Dimensions.* New York: HarperCollins, 2005.

Russell, Bertrand. *Wisdom of the West.* New York: Crescent Books, 1959.

Schwartz, Arthur (Music) and Dietz, Howard (Lyrics). "Dancing in the Dark" from *The Band Wagon.* Performed in New York, 1931. From the CD *Dancing In The Dark.* ASV Ltd., England, 2002.

Siegfried, Tom. *Undiscovered Ideas at the Frontiers of Space and Time.* New York: Berkley Books, 2002.

Smolin, Lee. *The Trouble with Physics*. Boston: Houghton Mifflin, 2007.

Stace, Walter T. *The Teachings of the Mystics*. New York: The New American Library, 1960.

Stewart, Ian. *Why Beauty Is Truth.* New York: Basic Books, 2007.

Underhill, Evelyn. *Mysticism.* London: Methuen & Co. Ltd., 1960.

Vaughn, R. A. *Hours with the Mystics*, Vol. 1. London: Strahan and Company Ltd., 1879.

Von Hugel, Friedrich. *The Mystical Element of Religion*, Vol. 1. London: James Clark & Co. Ltd., 1961.

Wright, Karen. *The Master's Mistakes, Discover Magazine*, Special Einstein Edition, September, 2002. 50–53.

Zaehner, R. C. *Mysticism: Sacred and Profane.* Oxford University Press, New York, 1961.

Index

N

Q